Information and the Professional Scientist and Engineer

Information and the Professional Scientist and Engineer has been co-published simultaneously as *Science & Technology Libraries*, Volume 21, Numbers 3/4 2001.

History of Science and Technology: A Sampler of Centers and Collections of Distinction, edited by Cynthia Steinke, MS (Vol. 14, No. 4, 1995). *"A 'grand tour' of history of science and technology collections that is of great interest to scholars, students and librarians." (Jay K. Lucker, AB, MSLS, Director of Libraries, Massachusetts Institute of Technology; Lecturer in Science and Technology, Simmons College, Graduate School of Library and Information Science)*

Instruction for Information Access in Sci-Tech Libraries, edited by Cynthia Steinke, MS (Vol. 14, No. 2, 1994). *"A refreshing mix of user education programs and contain[s] many examples of good practice." (Library Review and Reference Reviews)*

Scientific and Clinical Literature for the Decade of the Brain, edited by Tony Stankus, MLS (Vol. 13, No. 3/4, 1993). *"This format combined with selected book and journal title lists is very convenient for life science, social science, or general reference librarians/bibliographers who wish to review the area or get up to speed quickly." (Ruth Lewis, MLS, Biology Librarian, Washington University, St. Louis, Missouri)*

Sci-Tech Libraries of the Future, edited by Cynthia Steinke, MS (Vol. 12, No. 4 and Vol. 13, No. 1, 1993). *"Very timely. . . . Will be of interest to all libraries confronted with changes in technology, information formats, and user expectations." (LA Record)*

Science Librarianship at America's Liberal Arts Colleges: Working Librarians Tell Their Stories, edited by Tony Stankus, MLS (Vol. 12, No. 3, 1992). *"For those teetering on the tightrope between the needs and desires of science faculty and liberal arts librarianship, this book brings a sense of balance." (Teresa R. Faust, MLS, Science Reference Librarian, Wake Forest University)*

Biographies of Scientists for Sci-Tech Libraries: Adding Faces to the Facts, edited by Tony Stankus, MLS (Vol. 11, No. 4, 1992). *"A guide to biographies of scientists from a wide variety of scientific fields, identifying titles that reveal the personality of the biographee as well as contributions to his/her field." (Sci Tech Book News)*

Information Seeking and Communicating Behavior of Scientists and Engineers, edited by Cynthia Steinke, MS (Vol. 11, No. 3, 1991). *"Unequivocally recommended. . . . The subject is one of importance to most university libraries, which are actively engaged in addressing user needs as a framework for library services." (New Library World)*

Technology Transfer: The Role of the Sci-Tech Librarian, edited by Cynthia Steinke, MS (Vol. 11, No. 2, 1991). *"Educates the reader about the role of information professionals in the multifaceted technology transfer process." (Journal of Chemical Information and Computer Sciences)*

Electronic Information Systems in Sci-Tech Libraries, edited by Cynthia Steinke, MS (Vol. 11, No. 1, 1990). *"Serves to illustrate the possibilities for effective networking from any library/information facility to any other geographical point." (Library Journal)*

The Role of Trade Literature in Sci-Tech Libraries, edited by Ellis Mount, DLS (Vol. 10, No. 4, 1990). *"A highly useful resource to identify and discuss the subject of manufacturers' catalogs and their historical as well as practical value to the profession of librarianship. Dr. Mount has made an outstanding contribution." (Academic Library Book Review)*

Role of Standards in Sci-Tech Libraries, edited by Ellis Mount, DLS (Vol. 10, No. 3, 1990). *Required reading for any librarian who has been asked to identify standards and specifications.*

Relation of Sci-Tech Information to Environmental Studies, edited by Ellis Mount, DLS (Vol. 10, No. 2, 1990). *"A timely and important book that illustrates the nature and use of sci-tech information in relation to the environment." (The Bulletin of Science, Technology & Society)*

End-User Training for Sci-Tech Databases, edited by Ellis Mount, DLS (Vol. 10, No. 1, 1990). *"This is a timely publication for those of us involved in conducting online searches in special libraries where our users have a detailed knowledge of their subject areas." (Australian Library Review)*

Sci-Tech Archives and Manuscript Collections, edited by Ellis Mount, DLS (Vol. 9, No. 4, 1989). *Gain valuable information on the ways in which sci-tech archival material is being handled and preserved in various institutions and organizations.*

Collection Management in Sci-Tech Libraries, edited by Ellis
*"An interesting and timely survey of current issues in colle
science and technology libraries." (Barbara A. List, AMLS
Development, Science & Technology Research Center, and
Research Libraries, New York Public Library)*

The Role of Conference Literature in Sci-Tech Libraries, edit
2, 1989). *"The volume constitutes a valuable overview of th
users by one of the most frustrating and yet important sour
information." (Australian Library Review)*

Adaptation of Turnkey Computer Systems in Sci-Tech Librar
(Vol. 9, No. 1, 1989). *"Interesting and useful. . . . The book
associated with the installation of a turnkey or ready-made
technical library." (Information Retrieval & Library Autom*

Sci-Tech Libraries Serving Zoological Gardens, edited by Elli
*"Reviews the history and development of six major zoologi
(Australian Library Review)*

Libraries Serving Science-Oriented and Vocational High Sch
(Vol. 8, No. 3, 1989). *A wealth of information on the specia
and vocational high schools, with a look at their services, s*

Sci-Tech Library Networks Within Organizations, edited by E
1988). *Offers thorough descriptions of sci-tech library netw
a common sponsorship or ownership.*

One Hundred Years of Sci-Tech Libraries: A Brief History, e
No. 1, 1988). *"Should be read by all those considering, or
information retrieval, whether in Sci-tech libraries or other
& Technical Services)*

Alternative Careers in Sci-Tech Information Service, edited b
1987). *Here is an eye-opening look at alternative careers fo
background, including librarians, scientists, and engineers.*

Preservation and Conservation of Sci-Tech Materials, edited
1987). *"This cleverly coordinated work is essential reading
practicing librarians. . . . Recommended reading." (Science*

Sci-Tech Libraries Serving Societies and Institutions, edited b
1987). *"Of most interest to special librarians, providing the
libraries and their activities as well as a means of identifyin
collections which may be of use to them." (Sci-Tech Librari*

Innovations in Planning Facilities for Sci-Tech Libraries, edi
No. 1, 1986). *"Will prove invaluable to any librarian establ
contemplating expansion." (Australasian College Libraries,*

Role of Computers in Sci-Tech Libraries, edited by Ellis Moun
*very readable text. . . . I am including a number of the articl
(C. Bull, Kingstec Community College, Kentville, Nova Sco*

Weeding of Collections in Sci-Tech Libraries, edited by Ellis M
*"A useful publication. . . . Should be in every science and te
Library Review)*

Sci-Tech Libraries in Museums and Aquariums, edited by Ell
1985). *"Useful to libraries in museums and aquariums for i
information." (The Association for Information Managemen*

Data Manipulation in Sci-Tech Libraries, edited by Ellis Mou
*"Papers in this volume present evidence of the growing sop
data by information personnel." (Sci-Tech Book News)*

Role of Maps in Sci-Tech Libraries, edited by Ellis Mount, DL
*about the acquisition of maps and the special problems of th
insightful book.*

Fee-Based Services in Sci-Tech Libraries, edited by Ellis Mount, DLS (Vol. 5, No. 2, 1985). *"Highly recommended. Any librarian will find something of interest in this volume." (Australasian College Libraries)*

Serving End-Users in Sci-Tech Libraries, edited by Ellis Mount, DLS (Vol. 5, No. 1, 1984). *"Welcome and indeed interesting reading. . . . a useful acquisition for anyone starting out in one or more of the areas covered." (Australasian College Libraries)*

Management of Sci-Tech Libraries, edited by Ellis Mount, DLS (Vol. 4, No. 3/4, 1984). *Become better equipped to tackle difficult staffing, budgeting, and personnel challenges with this essential volume on managing different types of sci-tech libraries.*

Collection Development in Sci-Tech Libraries, edited by Ellis Mount, DLS (Vol. 4, No. 2, 1984). *"Well-written by authors who work in the field they are discussing. Should be of value to librarians whose collections cover a wide range of scientific and technical fields." (Library Acquisitions: Practice and Theory)*

Role of Serials in Sci-Tech Libraries, edited by Ellis Mount, DLS (Vol. 4, No. 1, 1983). *"Some interesting nuggets to offer dedicated serials librarians and users of scientific journal literature. . . . Outlines the direction of some major changes already occurring in scientific journal publishing and serials management." (Serials Review)*

Planning Facilities for Sci-Tech Libraries, edited by Ellis Mount, DLS (Vol. 3, No. 4, 1983). *"Will be of interest to special librarians who are contemplating the building of new facilities or the renovating and adaptation of existing facilities in the near future. . . . A useful manual based on actual experiences." (Sci-Tech News)*

Monographs in Sci-Tech Libraries, edited by Ellis Mount, DLS (Vol. 3, No. 3, 1983). *This insightful book addresses the present contributions monographs are making in sci-tech libraries as well as their probable role in the future.*

Role of Translations in Sci-Tech Libraries, edited by Ellis Mount, DLS (Vol. 3, No. 2, 1983). *"Good required reading in a course on special libraries in library school. It would also be useful to any librarian who handles the ordering of translations." (Sci-Tech News)*

Online versus Manual Searching in Sci-Tech Libraries, edited by Ellis Mount, DLS (Vol. 3, No. 1, 1982). *An authoritative volume that examines the role that manual searches play in academic, public, corporate, and hospital libraries.*

Document Delivery for Sci-Tech Libraries, edited by Ellis Mount, DLS (Vol. 2, No. 4, 1982). *Touches on important aspects of document delivery and the place each aspect holds in the overall scheme of things.*

Cataloging and Indexing for Sci-Tech Libraries, edited by Ellis Mount, DLS (Vol. 2, No. 3, 1982). *Diverse and authoritative views on the problems of cataloging and indexing in sci-tech libraries.*

Role of Patents in Sci-Tech Libraries, edited by Ellis Mount, DLS (Vol. 2, No. 2, 1982). *A fascinating look at the nature of patents and the complicated, ever-changing set of indexes and computerized databases devoted to facilitating the identification and retrieval of patents.*

Current Awareness Services in Sci-Tech Libraries, edited by Ellis Mount, DLS (Vol. 2, No. 1, 1982). *An interesting and comprehensive look at the many forms of current awareness services that sci-tech libraries offer.*

Role of Technical Reports in Sci-Tech Libraries, edited by Ellis Mount, DLS (Vol. 1, No. 4, 1982). *Recommended reading not only for science and technology librarians, this unique volume is specifically devoted to the analysis of problems, innovative practices, and advances relating to the control and servicing of technical reports.*

Training of Sci-Tech Librarians and Library Users, edited by Ellis Mount, DLS (Vol. 1, No. 3, 1981). *Here is a crucial overview of the current and future issues in the training of science and engineering librarians as well as instruction for users of these libraries.*

Networking in Sci-Tech Libraries and Information Centers, edited by Ellis Mount, DLS (Vol. 1, No. 2, 1981). *Here is an entire volume devoted to the topic of cooperative projects and library networks among sci-tech libraries.*

Planning for Online Search Service in Sci-Tech Libraries, edited by Ellis Mount, DLS (Vol. 1, No. 1, 1981). *Covers the most important issue to consider when planning for online search services.*

Information and the Professional Scientist and Engineer

Virginia Baldwin
Julie Hallmark
Editors

Information and the Professional Scientist and Engineer has been co-published simultaneously as *Science & Technology Libraries*, Volume 21, Numbers 3/4 2001.

NEW YORK AND LONDON

First Published by

The Haworth Information Press®,10 Alice Street, Bingha

Transferred to Digital Printing 2011 by Routledge
711 Third Avenue, New York, NY 10017
2 Park Square, Milton Park, Abingdon, Oxon, OX14 4RN

Information and the Professional Scientist a
co-published simultaneously as *Science & Tec*
21, Numbers 3/4 2001.

Cover design by Brooke R. Stiles.

Library of Congress Cataloging-in-

Information and the professional scientist and engineer / V
tors.

p. cm.

Co-published simultaneously as Science & Technolog

ISBN 0-7890-2162-5 (hard : alk. paper) – ISBN 0-789

1. Technical libraries. 2. Scientific libraries. 3. Librari
mation resources. 4. Electronic information resources–Use stu
ing. 6. Science–Information services. 7. Technology–Inform
Hallmark, Julie. III. Science & technology libraries.
Z675.T3I455 2003
026'.6–dc21

Publisher's Note

The publisher has gone to great lengths to ensure the quali
but points out that some imperfections in the original may

Information and the Professional Scientist and Engineer

CONTENTS

Introduction 1
Virginia Baldwin

Conversations with Chemists: Information-Seeking Behavior of Chemistry Faculty in the Electronic Age 5
David Flaxbart

Finding Physical Properties of Chemicals: A Practical Guide for Scientists, Engineers, and Librarians 27
A. Ben Wagner

Information-Seeking and Communication Behavior of Petroleum Geologists 47
Lura E. Joseph

Online Bibliographic Sources in Hydrology 63
Emily C. Wild
W. Michael Havener

The Latest and the Best: Information Needs of Pharmacists 87
Mignon Adams

The San Diego Zoo Library "Began with a Roar" 101
Linda L. Coates

Botanical Information: Resources and User Needs 121
Susan Fraser

Distinguishing Engineers from Scientists–The Case for an Engineering Knowledge Community 131
Thomas E. Pinelli

Current Awareness Reports at Albany International Research Co. 165
Alex Caracuzzo

Supporting the Information Needs of Geogr
Information Systems (GIS) Users in an A
Julie Sweetkind-Singer
Meredith Williams

Interdisciplinary Research: A Literature-Ba
of Disciplinary Intersections Using a Co
Geographic Information System (GIS)
Robert S. Allen

Index

ABOUT THE EDITORS

Virginia Baldwin, MS, MLS, is Associate Professor and Head of the Engineering Library and the Physics and Astronomy Library at the University of Nebraska in Lincoln. She is the liaison from the Sci-Tech Division of the Special Libraries Association to the Sci-Tech Section of the Association of College and Research Libraries and is a member of both organizations. She is also a member of the American Society for Engineering Education. Ms. Baldwin's work has appeared in various publications, including *Electronic Collection Management* (The Haworth Press, Inc.), *College and Research Libraries, Illinois Libraries,* and the *Journal of Technology Studies.*

Julie Hallmark, PhD, is Professor in the School of Information at the University of Texas in Austin. She is former President of the Geoscience Information Society and has held several offices in the Special Libraries Association at the state and national levels. Dr. Hallmark has been published in *Electronic Library,* the *Journal of Education for Library and Information Science,* the *Journal of the American Society for Information Science, Nature, Biochemistry, Analytical Chemistry, Special Libraries, Science & Technology Libraries,* and *Information Technology and Libraries.*

Introduction

Engineers and scientists seek information in a variety of ways and many studies have been devised and published that investigated how they seek, use, and communicate information vital to their research and development activities. Electronic resources and methods of communication have caused change in all aspects of their behaviors relative to information. The articles in this work represent the knowledge of user behavior accumulated by librarians in academic, corporate, government, and organization libraries.

The authors who contributed to this volume discuss these issues and make distinctions among various science and engineering fields, among various types of libraries, among communities of users and among individuals within a field or discipline. Within these articles are insights into differences in commercial versus academic databases, personal or group subscriptions to database services as well as print or online journals that researchers may have above and beyond what the library has purchased. Discussions are relevant to liaison, collection development and instruction librarians, to faculty in library and information science programs, and to scientists and engineers for the results themselves as well as for the processes as models for gleaning client and patron information. This work is valuable for librarians who do library instruction in academic libraries and provides insight into teaching to student needs during their academic life and in their professional lives to come.

This work covers several specific fields: geology, chemistry, geology-hydrology, pharmacy, veterinary medicine, zoology, botany, and engineering. The articles give insight into the differences between the information-seeking behaviors of scientists, as compared to those of engineers. Finally, Geographic Information System (GIS) users themselves are addressed, and a study of the interconnectedness of disciplines associated with GIS, through publication, yields valuable insights concerning user needs for information in related disciplines.

[Haworth co-indexing entry note]: "Introduction." Baldwin, Virginia. Co-published simultaneously in *Science & Technology Libraries* (The Haworth Information Press, an imprint of The Haworth Press, Inc.) Vol. 21, No. 3/4, 2001, pp. 1-3; and: *Information and the Professional Scientist and Engineer* (ed: Virginia Baldwin, and Julie Hallmark) The Haworth Information Press, an imprint of The Haworth Press, Inc., 2001, pp. 1-3. Single or multiple copies of this article are available for a fee from The Haworth Document Delivery Service [1-800-HAWORTH, 9:00 a.m. - 5:00 p.m. (EST). E-mail address: docdelivery@haworthpress.com].

http://www.haworthpress.com/store/product.asp?sku=J122

10.1300/J122v21n03_01

In "Conversations with Chemists: Informatic istry Faculty in the Electronic Age," David Flax ble one-on-one interview approach to obtain major fields of chemistry: biochemistry, org chemistry. The interviews were based on five c sults provided in-depth and current information search chemists, seek, use and keep current w what tools they use, and how electronic journal Keeping up with the literature is very time-co have devised a variety of responses to accompli what some may find to be surprising comments

A. Ben Wagner, a science librarian in an a ployed as a chemist, combines knowledge of b Properties of Chemicals: A Practical Guide for brarians." Useful for the scientist or engineer these properties of chemicals, as well as for ac this article delineates what information is fre Wagner describes and evaluates these resourc with commercially published databases for te evaluated resources are some pertinent academ

Two articles relate to geology, one, from ar from a government librarian at a U.S. Geologic tion-Seeking and Communication Behavior of P seph compares and contrasts the information beh engineers, and researchers and traces the chang resources. Among the valuable insights in this a the current methods of information transfer, ins tween the communication of information and t and some of the current threats to geology in graphic Sources in Hydrology," Emily Wild and in great depth the many and varied government tion, at the state, regional, and federal levels, and Web sites. The authors describe the most effect this field to obtain information.

The next three articles, those by Mignon Ad Fraser, deal with specific fields: pharmacy, zool the information needs and behaviors of profe discuss information resources in these fields as erinary medicine, and the plant science interdis economic botanists, plant taxonomists, historians, geneticists, and biologists.

Thomas Pinelli coalesces previously published literature, including his own published research on the subject, in "Distinguishing Engineers from Scientists–The Case for an Engineering Knowledge Community." In this literature review and in the results of Pinelli's own studies in the aerospace industry is a vast synopsis of the information-seeking and use behaviors of scientists and engineers, and comparisons and contrasts of science and technology and of scientists and engineers. The result is especially valuable for the librarian who serves to provide and instruct regarding the information needs of the engineering profession, professor, and student.

Adding to our insight into the information needs and behaviors of engineers is Alex Caracuzzo's article "Current Awareness Reports at Albany International Research Co." In this article, a librarian in a corporate library describes how he addresses the information needs of engineers and researchers in a specific engineering field, engineered textiles. In this article are informative descriptions of categories of information needs and some surprising resources for fulfilling them.

The remaining two articles deal with the highly interdisciplinary research tool and field of study, Geographic Information Systems (GIS). In "Supporting the Information Needs of Geographic Information Systems (GIS) Users in an Academic Library," Julie Sweetkind-Singer and Meredith Williams describe Stanford University Library System's solution to making their GIS access flexible, well managed, highly supported, and publicized for the myriad of uses, research studies and discipline support that are applicable to this system. Robert Allen used the GIS as the common tool for his research into discipline usage of GIS, and into the overlap and connectivity of disciplines related to GIS through publications. In his article, "Interdisciplinary Research: A Literature-Based Examination of Disciplinary Intersections Using a Common Tool, Geographic Information System (GIS)," Allen related, among other findings, that the highest interdisciplinarity among the eighteen disciplines that were interconnected in the literature examined occurred for transportation, engineering, environmental science, and hydrology.

Knowledge about the information needs of scientists and engineers, as professionals in the many fields covered in this work, gives insight into the specific educational needs of college and university students. Use of this information will be an asset to librarians and departmental faculty in including appropriate instruction for science and engineering classes.

Virginia Baldwin

Conversations with Chemists: Information-Seeking Behavior of Chemistry Faculty in the Electronic Age

David Flaxbart

SUMMARY. Six faculty members in the Department of Chemistry and Biochemistry at the University of Texas at Austin were interviewed one-on-one to gather information about their information-seeking behavior, favored resources, and opinions about the transition from a print to an electronic information environment. In most cases, these chemistry faculty members have eagerly embraced the enhanced access to chemical information made possible by the steady addition of electronic journals and networked database systems. The most-cited benefits include significant time-saving and convenience as well as access to more journals than ever. As a result, use of the physical library and its printed collections by faculty is declining. Chemistry faculty interviewed expressed a strong self-reliance in their information-seeking skills and showed sophistication in their choice of tools.

KEYWORDS. Information-seeking behavior, electronic journals, chemists, chemistry faculty

---

David Flaxbart, MLS, is Head, Mallet Chemistry Library, University of Texas at Austin, Austin, TX (E-mail: flaxbart@uts.cc.utexas.edu).

[Haworth co-indexing entry note]: "Conversations with Chemists: Information-Seeking Behavior of Chemistry Faculty in the Electronic Age." Flaxbart, David. Co-published simultaneously in *Science & Technology Libraries* (The Haworth Information Press, an imprint of The Haworth Press, Inc.) Vol. 21, No. 3/4, 2001, pp. 5-26; and: *Information and the Professional Scientist and Engineer* (ed: Virginia Baldwin, and Julie Hallmark) The Haworth Information Press, an imprint of The Haworth Press, Inc., 2001, pp. 5-26. Single or multiple copies of this article are available for a fee from The Haworth Document Delivery Service [1-800-HAWORTH, 9:00 a.m. - 5:00 p.m. (EST). E-mail address: docdelivery@haworthpress.com].

http://www.haworthpress.com/store/product.asp?sku=J122

10.1300/J122v21n03_02

INTRODUCTION

Understanding how scientists gather and use i important first step in developing library colle many facets: index searching, current awareness cation among faculty and between instructors ar library. The comparatively recent addition of di this topic even more challenging by adding con the behavior outside the library walls. Electronic cess, has generally increased the consumption o time provides different avenues for obtaining it.

A good portion of the literature on the "info scientists has focused on the usage of scientific jc of Donald King and Carol Tenopir is probably b been extensively cited.[1-2] Research on the beha group is somewhat less plentiful.[3-4] The adve sparked significant changes both in the way scie the way it is published.[5-7]

The predominant methodology in user-studie ologies used to examine user behavior include discussions,[8] and one-on-one interviews, the me

User-behavior studies, no matter how well ex inherent drawbacks. First, a user study dates very information services is currently so rapid that it i snapshot of it, and that snapshot may soon be of s parison of studies over time, to detect trends an cause of differing methodologies and subject gro some studies to statistically over-analyze sparse

Further, there can be a significant disconnect asked in surveys, the respondents' understandi investigator's interpretation of the responses. somewhat differently, and much can be lost view-based study can circumvent this problem adjust and adapt a set of questions for each resp on the spot, and ask follow-up questions to elicit drawback is that interviews take much more tim is much smaller than a survey can generate, thus results impractical.

Finally, any study that focuses on a single in cult to extrapolate reliably to other populations in available resources and services among insti

So why do this? What can be gained from studying information-seeking behavior among faculty? First and foremost, any excuse to sit down with one's faculty and discuss the information landscape is a good one. It helps educate the faculty, inform the librarian, and reduce the distance between them. In this sense individual interviews are much more meaningful than surveys.

Second, gathering details about how the faculty use library services, or don't use them as the case may be, can be useful both in validating the choices that the library has made recently, as well as inform future choices. Not only does this influence the selection (and deselection) of resources, it can also extend to the design of the library Web site, and the development of new services and outreach methods.

Finally, despite the transitory nature of user studies, some general conclusions can be reached that can and do extend beyond the immediate future. Some of them may even contradict common wisdom about what faculty do and how they do it, and break down a few stereotypes about scientists in general and chemists in particular.

CHEMISTS AND THEIR LIBRARIES

Chemistry is a highly collaborative science, whose core functioning unit is the research group. Laboratory work is carried out largely by graduate students and postdocs. The professor organizes and directs the work, prepares grant applications, administers the funding, and supervises the workers. Papers to be submitted for publication might be written by the professor, but may also be drafted by the person who has done most of the lab work, who will usually be the lead author of the paper. The graduate students are often engaged in two-pronged workloads: participation in the team's work as a whole and their own research for their dissertations. In general these two directions are intertwined and often cannot be easily separated. There is frequently collaboration among different research groups, either within the same institution or with others elsewhere. Interdisciplinary collaboration is also of growing importance, and research efforts frequently cross traditional departmental lines.

The work of a chemistry professor is multifaceted and extremely time-consuming. In addition to regular teaching and committee duties, the professor must oversee a group that might be quite large. This administrative role involves obtaining and dispersing funding, obtaining equipment, carrying out myriad administrative tasks, mentoring students, supervising dissertation research, and attending many meetings. An assistant professor seeking tenure must juggle these jobs–often without much experience, guidance, or administrative support or funding–and still find time to excel and make a name for

himself or herself in an extremely large and c
demic community. This can involve long workw
ligations can pile up because the new professor n
of senior colleagues everywhere, making it diffi
professor can be very stressful.

Knowing one's way around the scientific liter
many of these duties effectively. If an army trave
ists surely travel on their journals. The cutting e
chemical research are both found almost excl
peer-reviewed journals. Chemists have never v
such as preprints or conference proceedings, an
ment of the chemical literature, are not used as ir
chemists in industry.

Thus chemists are highly dependent on timely
journals in their field, which include rapid-comm
full paper journals, and review journals.

Most other things in the library–databases, m
ence books–are secondary in importance and are
journal articles. So chemists will ultimately judge
lection. In the past this meant print journal subscrip
obtained via interlibrary loan. Today this primarily
cess to electronic journals on the Web. As the avail
tal information formats accelerate, user expecta
challenging a library's ability to keep up both tech

And keeping up is the name of the game. Before
for a chemist to read and know almost all of the
plus a fair amount of that outside his immediat
course, no longer even remotely possible. Chemist
read the most crucial publications for their researc
relentlessly. Yet a person does not become a chem
read about the research of others–one's own resear
the ability to search and gather the literature quick
tant. Every hour spent reading or photocopying is
ing and writing.

Much of their professional lives is wrapped u
communication: writing and submitting papers, r
als by others, editing journals, browsing, reading,
cles, maintaining personal reprint files, and searc
The literature can fascinate, annoy, bore, surpris
it is just a means to an end. That end is the creatic
as the cycle repeats itself, creates more literature

THE SETTING

The University of Texas at Austin (UT) is a large public institution (over 52,000 students, over 11,000 of which are graduate students) with major programs in science and engineering. UT is the fourth-largest producer of Ph.D.s in the United States. Due to odd twists of history, the Austin campus has a large College of Pharmacy and a nursing school but does not have a medical school or a medical library. It is the flagship campus of the University of Texas System, which has eight other academic campuses and six medical institutions scattered around the state.

The Department of Chemistry and Biochemistry, a unit of the College of Natural Sciences, has over fifty tenure-track faculty positions and active interdisciplinary collaborations with a number of other academic departments, colleges, and research institutes on the Austin campus. Approximately 270 graduate students were enrolled in the fall of 2002, almost all of whom are on Ph.D. tracks. In addition, the department employs nearly one hundred postdoctoral researchers and other research scientists in various capacities, as well as around fifteen non-tenure-track instructors. There are 700-800 undergraduate chemistry and biochemistry majors; the latter outnumber the former. The Department occupies all of one of the largest buildings on the Austin campus, and also has labs in a number of other facilities nearby. UT-Austin spent over $11 million in chemical R&D funds in 1999.

The Chemistry Library is located in the center of the chemistry building on the ground floor. It is one of five science branch libraries on the main campus, which are administratively part of the UT General Libraries system. The Chemistry Library is responsible for collecting materials in chemistry, biochemistry, chemical engineering, and human nutrition and food science. While the Chemistry Library serves most of the needs of all chemists on campus, biochemists, medicinal chemists, and physical chemists also make extensive use of other branch libraries that hold materials they need. Founded at the same time as the University, in 1883, the Chemistry Library's collection is deep, with journal holdings extending back into the mid-19th century. Its collection numbers over 85,000 volumes, with over 300 current journal subscriptions. Approximately a quarter of the collection is currently housed in an offsite storage facility. The total annual budget for chemical information is around $700,000, which includes subscriptions to databases such as *SciFinder Scholar*, *Beilstein Crossfire*, and *Chemical Abstracts–Student Edition*.

UT-Austin has invested heavily in the future of online scholarly information. Due to budgetary necessity, the General Libraries began in 2000 to cancel print for many hundreds of journals, especially those from large commercial publishers such as Elsevier, Academic Press, Kluwer, and Wiley. The Chemistry Library has over one hundred such online-or continues to maintain print subscriptions to co Chemical Society, the Royal Society of Chemis Society, among others.

METHODOLOG

Six faculty members in the Dept. of Chemis lected and asked to participate in a personal inte utes. While these professors certainly do not sample, and were not chosen randomly, they w talking with a suitably representative cross-sect terviewees came from each of the three largest s partment: Biochemistry, Organic Chemistry, from each division was a "senior" tenured facul nior" tenure-track assistant professor, in order t tional differences among faculty. The one-on-c in the faculty members' offices during the sum

All faculty interviewed supervise active grou a number of graduate students, postdocs, and in search assistants. The six interviewees and thei will be designated thusly throughout the article

- Senior Biochemist: regulation and organi
- Junior Biochemist: bioinformatics of prot
- Senior Organic Chemist: natural product s teraction and enzyme mechanism
- Junior Organic Chemist: catalytic proces sis; nanostructured materials design and a
- Senior Physical Chemist: photophysica self-assembling polymers; polymer synth
- Junior Physical Chemist: spectroscopy neous materials

All interviews were based on a set of prepared sions ranged widely and occasionally went off tions fell into several distinct categories:

1. Background questions on who within the formation-gathering tasks

2. Tools that are used for these tasks (database selection and awareness)
3. Tools and techniques used for current awareness
4. The impact of electronic journals on research
5. The future of chemical information and science libraries

The questions were primarily designed with the intent of launching an engaging conversation, and to obtain a glimpse into the world of the working research group.

INTERVIEW RESULTS

Despite the wide range of interests represented in this group, the interviews drew some consistent responses and demonstrated that some generalizations about chemical information-gathering practices are valid.

Faculty Roles in Information Seeking

There has often been a vague assumption in libraries that science faculty do not carry out their own literature searching and information gathering–that others, especially graduate students, do it for them. While this may be true for rote in-library tasks such as retrieving and copying articles, the interviews indicated clearly that faculty do most or all of their own literature searching themselves. All but one indicated that they rely on their own skills to gather necessary information. The senior biochemist stated it best: "I can cast a wider net with the computer [than students can], and use my judgment for the decision on which ones to pursue. That's not left to the students."

The exception was the senior organic chemist, who said that he relies on students to do searches for projects currently underway, while he tends to do searching for the future directions of research. Partly this is due to time constraints, but it is also a pedagogical issue, since he feels students need to learn how to gather information on their own.

Most of this work is done in their offices, during "regular" working hours. (Chemists can keep rather long hours, which frequently include late evenings and weekends.) Work at home is often limited by the quality of equipment and internet connections, which are generally slower than ethernet connections on campus. The fact that major resources such as *SciFinder Scholar* and *Beilstein Crossfire* cannot be proxied for use via third-party Internet providers is also a factor, and thus home use is often limited to downloading and reading journal articles.[9] The junior biochemist said that he frequently downloads article PDFs in his office onto his laptop, then takes them elsewhere to read offline.

Each faculty member was asked to assess skills on a scale from "expert" to "good" to "ment." While modestly stating that there's al most rated themselves good or expert, and the other questions largely backed up this assessme said he is "adequate plus." The junior physical c good, in comparison to my students, who are bo that he can always find a needed paper first, an knowledge of the field as well as skills in selec bases–skills that come with experience.

Another question dealt with the flow of in group–who informs and instructs whom in the a tools. Again, most respondents indicated that a t Learning about new resources can often be ha email and paper mail descending on today's sc filters to make it manageable, and sometimes li do not make it through the first time. Viewin pages is a rare occurrence. Students and collea tidbits of news and advice, but it seems that fac to stay up with the newest developments, and

The junior physical chemist said that student proach' to finding literature, getting something it, rather than looking at much literature in gen from the old model of library browsing–a "just i losophy. The senior organic chemist indicated th assistance than the others do, asking them for he needs it. His students work more frequently wi *Crossfire*, and he finds it faster to ask them for himself.

However, none of the faculty stated that they in database use. Faculty may be more liable to p of relevant articles, PDF files, and email alert journal searching skills are self-taught. Students the library, but self-instruction is the norm, an point of need.

The Tools of the Trade

All the interviewees were asked about the pri use in information gathering: which one is the

what others they also use. The variety of responses here was definitely interesting. Somewhat surprisingly, *SciFinder Scholar* (Chemical Abstracts) is not always the tool of choice for some chemists.

The organic chemists benefit from having two excellent resources at their disposal. Both indicated frequent use of *SciFinder Scholar*, but also of *Beilstein Crossfire*, and said that the type of information need governed the choice between them.[10] The junior organic chemist prefers *Beilstein* for seeking preparation and reaction information, especially for natural product fragments, since this tool's structure searching capabilities are more powerful, and faster, than *SciFinder*'s. Topical and keyword searches, on the other hand, are much more effectively done in *SciFinder*. He indicated that both tools are used daily. The senior organic chemist likewise said that "*Beilstein* is a better tool for things we typically do," but that *SciFinder* was used slightly more than *Beilstein* because of its author and keyword capabilities.

The junior physical chemist also uses *SciFinder* as his primary tool, because of its good coverage of the physics literature. One drawback he described is *SciFinder*'s inability to do precise multi-field searches up front. (*SciFinder* requires the user to execute a single initial query by topic or author or chemical identifier, then refine or analyze the results after the fact.)

The senior physical chemist was the only one who indicated a preference for *Web of Science*, the public interface to ISI's *Science Citation Index*. He uses this tool daily because "it cuts across things in a unique way and avoids keyword problems by tracking citations and co-citations over time."

The two biochemists described the most sophisticated approaches to using databases to locate pertinent literature. Both biochemists indicated that *Medline* (or its free version, PubMed) is their primary tool for literature searching. This may be the result of individual research emphasis: the senior biochemist categorized himself on the biological, rather than the chemical, end of biochemistry. He described a highly specialized method of searching. His research group relies on EndNote bibliographic management software as a front end to *Medline*, simultaneously building a local database of bibliographic records for subsequent use in writing papers. He said that EndNote is "on all the time" on his computer, and that he searches both *Medline* and his local database several times a day at least.[11]

The junior biochemist, whose specialty involves bioinformatics, called *Medline* the most important tool in his field, and said it was used almost exclusively. While he uses the standard public interfaces to *Medline* (either Ovid or PubMed), his group also licenses the *Medline* database directly and loads it onto a local server. This file is then "mined" using custom-written Unix programs that look for articles reporting highly specific information on protein interactions. These searches primarily use the Abstract fields, rather than MeSH headings, which he characterized as "horrible, a indicated that more than half of the group's lite this data-mining technique, and that *Medline* is this purpose because of its size and age.

The faculty also described a variety of second complement the primary favorites. The senior or as an occasional backup to *SciFinder* and *Be* searches within specific publisher e-journal sites, cal Society's Web Editions and Elsevier's *Scienc* are usually limited to looking for papers by speci chemist indicated *Chemical Abstracts-Student E* as a favorite alternative due to its flexible multi-f also uses *Web of Science* when looking for the " topic, especially when he's engaged in writing a p he doesn't use it nearly as frequently as *SciFinder* times. Both physical chemists also listed *Chem* and *Web of Science* as alternatives for the same

The biochemists again displayed more ind tools to complement *Medline*. The junior bioch *Faculty of 1000* (Biology Reports Ltd.), *BioMe* known as NEC ResearchIndex, for computer sci *Science*. The mention of *Faculty of 1000* was p ternative, peer-recommended approach to the b

The senior biochemist mentioned tools such databases, that provide crucial genetic data in however that there is currently a tendency to priv and charge for access to them, which in most c bothers with them. (A specific example is the *Y* owned by InCyte.) He suspects that many of thes will fail in the near term for lack of subscribers. genuity to fall back on: "We're cheap, and we selves." The senior organic chemist also noted and various NIH sites as important resources.[13]

Finally, faculty were asked about their know free Web resources that are often linked to fro these are literature-based, while others are focu data. Surprisingly, some of these sites tended to terviewed, or known only by name, and rarely i matrix of these sites and the responses to them.

TABLE 1. Use of Selected Web-Based Resources

Resource	Sr. Org.	Jr. Org.	Sr. Phys.	Jr. Phys.	Sr. Bio-chem.	Jr. Bio-chem.
ChemWeb	O	N	N	U	N	H
ChemFinder	N	H	N	N	U	N
NIST Chemistry WebBook	N	N	N	U	O	N
BioMed Central	N	H	N	N	H	U
PubMed Central	H	N	N	N	U	U
PubScience (DOE)	N	N	N	N	N	N
sciBase	N	N	N	N	N	N
arXiv.org (LANL/Cornell)	N	N	N	O	N	N
sigmaaldrich.com	H	O	H	U	U	H

Key: U–Uses this resource; O–has used it occasionally but not regularly; H–has heard of it but doesn't use; N–has never used it.

Staying Up to Date

One of the most daunting problems facing today's scientists is keeping up with the literature in one's field. As the scientific literature has expanded relentlessly in past decades with more authors publishing more papers in more journals than ever before, the problem is not a minor one. It requires a concerted effort on the part of a scientist to stay abreast of the latest developments in even a highly specialized field, and this leaves little time for reading in "outside" areas.

Brown described preferred methods for staying up to date among chemists and biochemists: scanning current journal issues was far and away the most professed method.[14] While the tools, access methods and formats have all evolved, this is still largely true today. Fernandez conducted a survey that found that faculty still tend to use a wide variety of mechanisms to stay current in their literature, with table-of-contents (TOC) scanning being the most popular.[15]

Publishers have long acknowledged the importance of scanning tables of contents of journal issues, and many have set up free email services that will alert recipients when new issues of specific journals are published. Some messages contain that issue's table of contents; others merely include a URL where the TOC can be found on the Web. A major drawback to this type of service is that many journals are so large that TOCs are of marginal use to chemists looking for highly specific articles. Titles such as *Proceedings of the National Acad-*

emy of Sciences and *Journal of Biological Che* scanning can be quite time-consuming and often vices originate with the publisher and do not invo libraries have in turn offered locally-developed a

In the past, scanning journals required regular examining whatever personally subscribed journ braries provide a current periodicals area where A faculty member who set aside a particular time ing there could expect to stay up to date on intere scenario, however, is rapidly becoming obsolete interviewed for this article indicated that they rar this reason anymore; some never did in the first p chemist expressed a preference for browsing nev Unfortunately, he can no longer be assured of s cause the UT library has dropped so many print online counterparts over the last couple of year doubt continue due to ongoing budgetary constr

The junior biochemist subscribes to a numbe junior physical chemist avoids TOC-alerts becau large volumes of unnecessary email. He prefers t indexes like *SciFinder* and find new materia bles-of-contents on occasion, but this is not a pri

The senior organic chemist, as mentioned abo graphical table-of-contents (GTOCs) of journals browses them online when available. The junior ticulous written list of his core journals, marking ined online, so that he doesn't miss any. He als online, which maintains a degree of serendipity on keyword searching in indexes. Graphical tab abstracts), which have existed in print for many journal scanning in organic chemistry: they enab issue's contents looking not for key terms, but fo the substances and reactions central to each ar such as *Journal of Organic Chemistry*, *Tetrahed* *Organic Letters* have incorporated GTOCs int varying degrees of effectiveness. Special current as *Methods in Organic Synthesis* and *Natural Pr* Royal Society of Chemistry) are composed enti lected from other journals.

Third-party table-of-contents services have e Current Contents and UnCover are probably two

UT Libraries have made use of UnCover (now taken over by Ingenta) and subscribe to its Reveal alerting service, there is little evidence that it is used much by science faculty. UT has not subscribed to any Current Contents databases in the past, so these are not a local option. Furthermore, none of the faculty mentioned using the TOC viewing function within *SciFinder Scholar.* (Since this is a relatively new addition to Scholar, they may not be aware of it, and usage statistics indicate that this option is rarely used at UT.) The commercial version of *SciFinder* provides a "Keep Me Posted" function for registered users, but this is not available in the academic version.

Of the six faculty interviewed, only the senior biochemist indicated that he uses an independent automated TOC service. His group, with two others, shares a subscription to a service called Reference Update (RU), now provided by ISI and similar to Current Contents in the biomedical sciences. Subscribers receive a weekly download of new records covering about 1,100 biomedical journals, that can be searched within special client software supplied by the vendor. (RU does not cover core chemistry topics and journals.) This enables the user to create and store search profiles that are used over and over. Although the service is not cheap, the biochemist said that cost is not a factor: "RU is easily customized as my focus changes. I would happily pay for it all myself, [since] there's no way to keep up with the literature without something like this."[17]

The senior biochemist is also one of the handful of faculty members who still receive an electronic SDI ("Selected Dissemination of Information") via the library. An automated search profile within the STN online interface is run against new records added to the CA file every two weeks, and results are emailed to him. The library picks up the charges for this service. This complements his other regular searching in *Medline* and Reference Update, and retrieves some items of more strictly chemical interest that would otherwise be missed. Due to the costs involved, this is not a service that is widely advertised to faculty.

Given the overall difficulties in scanning new journals for articles of interest and the apparent reluctance to use publisher-based or third-party alerting services, do these faculty feel that they're really keeping up to speed with the literature? Their responses to this question varied. The senior biochemist, who has by far the most elaborate literature retrieval setup, feels he is up to speed, despite the steadily increasing volume of literature. The junior biochemist speculates that he "misses a fair bit" and would like to see better alerting services to streamline the process, although time is a limiting factor that will never go away.

Both the junior and senior physical chemists confessed that they miss way too much. The senior one expressed "semi-angst" in admitting that he hasn't been able to keep up for years, though he's generally aware of developments in specific areas. When he moves in a new directi required, especially when a new grant proposa cite relevant research can cause delays in the gr potential embarrassment). The junior physical dencies to ratchet up searching and reading du time. He is often shocked by what he has missec the iceberg. But "as long as we're not complete make contributions."

In contrast, both organic chemists feel they ture, though they must be more selective in wh

A follow-up question asked the faculty men now read more or fewer articles, and scan more in the pre-online past. Again, the responses vari responded that he now reads and scans more th availability and convenience. The senior organ reckons he reads less. As he still relies on print, t tion of print of journals he used to scan. The juni that it's much easier to obtain, store, and forw reckons he has thousands squirreled away on his time to read them all. His reading volume pr though he is probably now selecting better thin believes he reads and scans more now, and citat nals and more papers in a given area that he wo scanning.

The junior biochemist stated that even thou articles, he has "time to look at a lot less." As a s ing fields of genomics and proteomics, where number of journals are growing quickly, he mu to. His senior colleague believes he is reading th cles, and scanning more journals, due also to the

Despite the increased access and convenien journals, it is obvious that the principal limitin ture consumption pattern is time. More article and flow across the desktop, and a 10-gigabyt number of PDF files, but only a finite number c lated. Selectivity is crucial: the chemist must most promising items to read, and these tend to prestigious "top" journals. While there are thou chemical and biochemical fields, it is very like consulted. Large e-journal package subscrip which bundle the good with the mediocre, ha lower-tier journals, few of which had wide print distribution. Both citation studies and analysis of consortial e-journal usage statistics show that their usage nevertheless remains low compared to that of top-tier titles.[18] Bradford's Law, it seems, is alive and well.

The Changes Brought by Electronic Journals

It is difficult to overstate the impact electronic journals have had on the practice of science. The profound nature of the changes brought about by desktop access to journals makes it difficult to believe that they have only been in existence for less than ten years, in many cases less than five years. The rate of adoption of new formats by both publishers and end-users has outstripped even the most optimistic estimates of the mid-1990s. No matter how one measures it–by levels of comfort, levels of usage, expressions of satisfaction–scientists have embraced electronic access with open arms, even though some remain troubled about long-term issues such as archival permanence, economics, and the serendipity factor.

Early pilot projects with online journals, such as the CORE project at Cornell University in the early 1990s, showed chemists voicing concerns about the viability of journals on a computer.[19] Most faculty did not anticipate rapid acceptance of this new format in place of comfortable printed journals that had remained largely unchanged for many decades. Nor did many librarians. The fact that two extraordinarily conservative and cautious cultures–academic science and scholarly publishing–adopted electronic formats so quickly is testimony to the power and attraction of digital access. There will be no going back.

The interviews for this article offered the opportunity to hear opinions about electronic journals and their significance. While the interviews covered a variety of subjects, the faculty tended to direct much of the conversation toward this topic, making it clear that much of their thinking on chemical information revolves around electronic journals.

When asked how electronic journals have affected their work, the faculty interviewed were unanimous in their opinion that e-journals have brought major changes. The change cited most often is the saving of time. The junior organic chemist said it best: "Hours spent in the library are now reduced to seconds online. [It's] that much easier to stay ahead of the curve." The junior biochemist recalled the "miserable success rate" he used to experience in using his university library as a graduate student, and how electronic access has made obtaining articles so much easier and more efficient.

As much as some scientists and librarians like to wax nostalgic about the happy and fruitful hours spent exploring the library stacks, the basic reality is that many scientists (and students) never enjoyed the task: it was something put off as long as possible or neglected altoget involved in doing library research in the printec and microscopically-printed indexes, searching ing down missing journal volumes, recalling b photocopiers–are likely reasons why library us brace an alternate mode of access. Using any lil maddening–experience at times, especially if c say that the digital library is inherently easier to nized, or more complete–but it is certainly *fast* and retrieve information without leaving one's c shortcomings as far as users are concerned. You looking for, but at least you didn't waste four h

Other questions addressed their current use o they felt about losing access to many printed jo being dropped in favor of online-only access. that they come to the actual library less now tha that the Chemistry Library is in the same buildi physical chemist says he still comes to the lib close to his office), but rarely can muster the e branches. Almost all of his journal articles come ever. Immediate desktop access to articles is to value: the more effort it takes to get a copy, the worthwhile. The junior organic chemist expre e-journal access, but said he would seek out wl format or location.

The senior faculty might be expected to mi their younger colleagues, and this was reflected organic chemist was alone in saying that he still time. He is still a frequent visitor to the library, t nals to scan and use due to the cancellations of distresses him. He is reluctant to migrate to e comes necessary I won't do it." The senior phy similar sentiments, saying that he missed the through journal issues: coming across the unex online world, but "it doesn't seem to happen mu library less nowadays, "because it's easier not

The rush to convert libraries to digital forn searchers, is clearly not applauded by everyone, not to assume that it is. It can be tempting print-lovers as hold-overs of an earlier age, fro But there are demonstrable drawbacks to an o visible in the well-documented reluctance of many people, young and old alike, to read anything from computer screens. A paper print-out is still the final destination for most articles, which may explain why PDF formats are still far more popular with users than HTML versions of the same content.[20]

The Future

The concluding set of questions asked the faculty to give an overall impression of today's chemical information landscape, whether they think it is getting better or worse, simpler or more complex, and what they think about the future of science libraries.

The senior organic and physical chemists were ambivalent about the changes they've seen so far, and where they're leading. The latter called the current situation "chaotic" in that there are now so many places a user must remember to look online for pertinent information. He is also concerned that availability of the gray literature will suffer as researchers focus only on journals that are accessible online. (The very term "gray literature" may now be expanding its meaning to include journals that are not online, which must surely be an ominous warning to their publishers.)

The senior organic chemist decried the time it takes to navigate through multiple Web pages to reach a desired article–a task that could be done faster by flipping rapidly through a print journal. He looks forward to increased hypertextual crosslinking among journals and indexes, especially using metadata applied to chemical structures within the text. But he said that overall the situation is "not yet at the point where we need it to be."

The others interviewed were generally enthusiastic about the direction chemical information is taking, and feel that it is now easier to identify and obtain information. The junior physical chemist believes that "it's immensely easier to find stuff now," as well as less time-consuming. He doesn't hesitate to do exploratory literature searches on a whim, just to see what's out there. The rest echoed this opinion.

When asked about the future of science libraries, and what they'd like to see developed in the next few years, faculty mentioned things that are being actively considered as next steps in the online information infrastructure. The junior physical chemist, expressing amazement at how little genuine content is out there on the Web, sees a role for libraries in filtering the gems from the dirt. The junior organic chemist had a desire for more complete journal backfiles online, with more extensive crosslinking among them, as well as electronic versions of key reference works popular among organic chemists.

The theme of seamless linking back and forth from indexes to fulltext was mentioned again and again. This would require greater cooperation among publishers than currently exists, as well as furt such as DOI and OpenURL, and wider imple bridge the gap between databases and local su advantage of digital-only article features woul static PDF versions, which at this time they see ity and printability of the PDF clone still tru SGML versions. The junior physical chemist tions and applets in an article are eye-catching, b raw data, which is what a scientist really wants

Some of the faculty are acutely aware of t stakeholders in scientific information, particul permanence and cost. Others are only vaguely a not yet given them much thought. The younger picious of publishers' motives and long-term c older ones, who have worked longer with publi Electronic journals have a more tenuous qualit brary shelves. Entire collections can vanish wi crash of a server. The junior physical chemist afraid that libraries might be "scammed" in t threaten to take away access to crucial informati can't be met. Libraries need guaranteed perman paid for, and preferably they should only pay on The junior biochemist is equally uncomfortab control over vital research information. He kno Public Library of Science, but must still subn journals due to the demands of tenure and pron

CONCLUSION

While academic chemists certainly share n noted that the chemists interviewed for this ar search university, whose main interests are foc and publication, and the training of graduate st having access to a large library system offering STM journals, fairly comprehensive monogra certainly not all) of the major database system censing. Their responses might not coincide n smaller institutions whose primary mission is faculty interviewed for this study were chosen tion of research chemists according to seniority

While one would expect a variety of approaches to information-seeking, their remarks provided strong evidence that electronic access is taking over more completely and more rapidly than anyone could have predicted a few years ago. Chemists have largely overcome their initial reluctance to use and depend on electronic journals. Faculty, far from being slow to adapt, are leading the way, continuing to direct their research groups' information-seeking in the new environment. While there is certainly two-way flow in the groups as faculty and graduate students learn from and teach each other in many informal ways, faculty resist depending on others for their information needs. The faculty's level of sophistication in seeking information should not be underestimated. They are creative, canny consumers and searchers.

It is also clear that, for the most part, faculty are using the physical library much less, even when their offices are nearby. The time-savings and convenience of online journals and databases enable faculty to consume more information in less time. Faculty are supportive of the library, but admit that they will visit the facility less if they don't have to. The key task for librarians in the future will be ensuring that users make the connection between all the marvelous online resources they use and the library, which still must select, pay for, organize, and promote them. The library retains its crucial role of intermediary, and this function is more important than ever as users face a confusing array of choices, and a wide variation in content quality on the Web. The faculty interviewed understand that role and express hope that it will continue to grow.

Librarians studying the information-seeking behavior of chemists should avoid focusing too much on particular tools and resources. When choices are available, chemists will choose tools that suit them best, and these vary according to subject specialty, type of need, and personal preference. *SciFinder Scholar* is an extremely broad and useful resource, but it is not the ideal tool for every information need in chemistry. Libraries can get tremendous mileage out of a few well-chosen (but often expensive) resources. Offering a variety of tools is important, along with the knowledge that nobody uses everything, but everybody uses something.

Chemists are happiest when they feel that their library is making genuine attempts to understand the uniqueness of their information resources and needs, and not lumping them together with other scientists. Chemistry is not like biology or medicine or physics. These fields certainly share similarities, but each has its own unique culture, vocabulary, and scholarly communication system. This caveat extends to the fact that Chemistry itself is not a homogeneous discipline. The same kinds of differences that separate broad disciplines also separate–to a somewhat lesser extent–subfields within the discipline.

As librarians are already well aware, the process of identifying documents has long since migrated to online database use. Printed indexes and alerting publications are relics of the past. Many academic libraries that have not already dropped subscriptions to the printed *Che* do so in the near future. However, the appropria *SciFinder Scholar*, is far from being affordabl some to retain printed CA against their prefere smaller schools, as well as those in developing Access to the tools of chemical information is a brarians and faculty together should continue to for all educational institutions.

While this study made no attempt to gather use, the opinions expressed by faculty point to of digital formats as the *primary* means of viewi search. As a new generation of graduate student very possible that their attitudes, coupled with eliminate the traditional printed scholarly jour most practicing chemists. Librarians will thus maintaining print subscriptions into the future. chival backfiles to digital formats will likewis off-site storage or even discarding altogether the further decreasing the use of physical libraries.

In some ways, though, the revolution is only h pabilities of the digital medium have made few journals. Features only possible in digital forma tures, animations of dynamical processes, raw tions, applets, metadata, and Chemical Marku waiting for wider adoption.[22]

Scientists and editors have welcomed the el for their convenience, power, and speed, but t scholarly journal and the articles within it has output is most often a print-out or photocopy printed page. Is this because the traditional printe mation package, that doesn't need any bells an and libraries can see the limitations of the two-di page, it is the authors, editors, and readers (who will have to lead the migration to more evolved

It remains to be seen whether electronic form scholarly communication as profound as the ong entists and librarians both have a very large sta work together to ensure that it is positive for ev ions of academic chemistry faculty matter a great fit enormously from keeping a finger on the puls consumers of chemical information.

NOTES

1. Tenopir, Carol and King, Donald W. 2000. *Towards electronic journals: realities for scientists, librarians and publishers*. (Washington DC: Special Libraries Association).

2. King, Donald W. and Tenopir, Carol. 1999. Using and reading scholarly literature. *Annual Review of Information Science and Technology (ARIST)* 34:423-77.

3. Williams, Ivor A. 1993. How chemists use the literature. *Learned Publishing* 6(2):7-14.

4. Brown, Cecilia. 1999. Information seeking behavior of scientists in the electronic information age: astronomers, chemists, mathematicians, and physicists. *Journal of the American Society for Information Science* 50(10):929-43.

5. Stewart, Linda. 1996. User acceptance of electronic journals: interviews with chemists at Cornell University. *College & Research Libraries* 57(4):339-49.

6. Stankus, Tony. 1999. The key trends emerging in the first decade of electronic journals in the sciences. *Science & Technology Libraries* 18(2-3):5-20.

7. Sathe, Nila A., Grady, Jenifer L., and Giuse, Nunzia B. 2002. Print versus electronic journals: a preliminary investigation into the effect of journal format on research processes. *Journal of the Medical Library Association* 90(2):235-43.

8. Young, Nancy J. and von Seggern, Marilyn. 2001. General information seeking in changing times. *Reference & User Services Quarterly* 41(2):159-69.

9. There are some possible technical solutions to this problem, but they have not as yet been investigated at the author's institution, for various practical and financial reasons.

10. *SciFinder Scholar* is the proprietary academic interface to the Chemical Abstracts' suite of databases. These include the CAPLUS file, Chemical Abstracts back to 1907; the Registry file, a database of chemical structures and names containing more than 40 million compound records; the CASREACT file, indexing several million organic reactions back to 1975; and subsidiary files such as CHEMLIST (regulatory information on chemicals) and CHEMCATS (chemical supplier catalogs). The Registry and CASREACT databases are searchable by chemical structure drawing as well as character-string queries. *SciFinder* is a site-licensed product of Chemical Abstracts Service, a division of the American Chemical Society. *Beilstein Crossfire* is a proprietary interface to the *Beilstein* database of organic structure and property information, containing records for over seven million carbon compounds and seven million organic reactions, with associated physicochemical data and literature references. *Beilstein Crossfire* is a product of MDL Information Systems, a subsidiary of Elsevier Science. It is made available for academic site licensing via the *Beilstein* Minerva Consortium in cooperation with the University of Wisconsin. Both *SciFinder Scholar* and *Beilstein Crossfire* require special client software to be installed on a user's machine.

11. EndNote allows direct connections to Z39.50-compliant databases and provides a generalized search interface to specific files. Records retrieved can be directly imported to a locally-maintained database and then searched later. A number of chemistry graduate students have indicated a preference for this type of searching interface, and express disappointment at the inability to connect to databases such as *Chemical Abstracts Student Edition* (FirstSearch) and *SciFinder*, which are not Z39.50-compliant. The library usually discourages this technique, since it bypasses many of the unique features of native database search interfaces. Records retrieved via native search interfaces can however be imported into bibliographic management software

with appropriate filters. EndNote is a product of the I (ISI). URL: http://www.endnote.com/.

12. Faculty of 1000: http://www.facultyof100 page: "Faculty of 1000 is a new online research tool t papers in biology, based on the recommendations (Viewed Sept. 5, 2002) It is available via personal or Central: http://www.biomedcentral.com/ CiteSeer: ht

13. National Human Genome Research Institute (gov/ InCyte Yeast Proteome Database: http://www databases/YPD.shtml Protein Data Bank: http://www.

14. Brown, "Information seeking behavior of sci

15. Fernandez, Leila. 2002. User Perceptions o Faculty Survey. *Issues in Science and Technology Li* http://www.istl.org/istl/02-winter/article3.html.

16. Schlembach, Mary C. 2001. Trends in curre *Technology Libraries* 20(2-3):121-32.

17. Reference Update: http://www.isinet.com/isi

18. Davis, Philip M. 2002. Patterns in electron composition of geographic consortia. *College & Res*

19. Stewart, "User acceptance of electronic jour

20. In 2001, Austin users downloaded 46,865 ar cal Society Web Editions system; 86% of these wer

21. Public Library of Science: http://www.publi

22. For example, the *Internet Journal of Chem* com/) began in 1998 with assistance from ARL/SPA capabilities of the new medium. But the number of arti in each successive year: 38 were published in 1998, tember).

Finding Physical Properties of Chemicals: A Practical Guide for Scientists, Engineers, and Librarians

A. Ben Wagner

SUMMARY. Free Internet resources that provide a significant amount of physical property information are critically reviewed. These free Web sites cannot totally replace classic print sources and online subscription databases. However, a number of these Web sites provide extensive and reliable property information, especially for substances that are used in commerce or have seen significant research interest. This guide will be of particular value to those with occasional needs for physical properties or working without the benefit of access to a major research library and its subscription-based electronic resources.

KEYWORDS. Physical properties, chemical compounds, Internet resources, World Wide Web guides, critical review, chemical information

INTRODUCTION AND METHODOLOGY

This article describes and evaluates the various Internet resources that provide significant physical property information and are available on the Web

---

A. Ben Wagner, BA, MLIS, is Sciences Librarian, University at Buffalo, The State University of New York (E-mail: abwagner@acsu.buffalo.edu).

[Haworth co-indexing entry note]: "Finding Physical Properties of Chemicals: A Practical Guide for Scientists, Engineers, and Librarians." Wagner, A. Ben. Co-published simultaneously in *Science & Technology Libraries* (The Haworth Information Press, an imprint of The Haworth Press, Inc.) Vol. 21, No. 3/4, 2001, pp. 27-45; and: *Information and the Professional Scientist and Engineer* (ed: Virginia Baldwin, and Julie Hallmark) The Haworth Information Press, an imprint of The Haworth Press, Inc., 2001, pp. 27-45. Single or multiple copies of this article are available for a fee from The Haworth Document Delivery Service [1-800-HAWORTH, 9:00 a.m. - 5:00 p.m. (EST). E-mail address: docdelivery@haworthpress.com].

http://www.haworthpress.com/store/product.asp?sku=J122

10.1300/J122v21n03_03

for free. Registration may be required, but no a attention was paid to sites providing data for a l large number of properties, or both. This review dividual substances. Excluded were sources spe ronmental information,[1-6] biomolecules[7-9] su atomic data,[13] and property estimation softw would be a major guide unto itself. Small, spe Web sites were also excluded, though some of are reviewed, especially in the polymer area. Th provide a more in-depth look at materials scien

A careful survey of existing science-oriented and Web directories was made to develop as con Included in this were academic Web guides a Chicago, Vanderbilt University, University at sity. Sources searched by *ChemFinder* (http:// reviewed as were standard directory guides from com), *Invisible Web* (http://www.invisibleweb. (http://dmoz.org/), and the Univ. of Wisc (http://scout.cs.wisc.edu/). Personal bookmarks chemical information work and listserv posting

THE IDEAL WOR

In an ideal world, all scientific libraries wou posal the many fine printed reference works a electronic databases covering physical proper printed sources are very old and out-of-print, would also have been in existence for decades a culties! A list of their holdings would include [K

- *Scifinder*® or *SciFinder*® *Scholar* [Chemi
- *Beilstein Handbook* (p) or *CrossFire*™ *Be* tems] (e)
- *Gmelin Handbook* (p) or *CrossFire*™ *Gr* tems] (e)
- *Landolt-Börnstein Numerical Data & Func* Verlag] (p,e)
- *International Critical Tables* [U.S. Natio
- The full spread of STN® International Da
- Standard Reference Data Program Public dards & Technology (NIST)] (p,e)

- Design Institute for Physical Property Data (**DIPPR**)® Publications (p,e)
- Thermodynamics Research Center (TRC) publications [Texas A&M Univ.] (p,e)
- Center for Information and Numerical Data Analysis and Synthesis (CINDAS) publications [Purdue Univ.] (p,e)
- *Handbook of Environmental Data on Organic Chemicals* [Wiley] (p)
- *Dictionary of Organic Compounds* and other Chapman Hall/CRC dictionaries (p,e)
- *Journal of Chemical and Engineering Data* [American Chemical Society] (p,e)
- *Journal of Physical and Chemical Reference Data* [Amer. Inst. of Physics/NIST] (p,e)

There are many reviews of the major print and electronic reference sources in individual articles and chemical information resource textbooks.[19-24]

STARTING POINTS

Manufacturer/Supplier Web Sites

Most chemical companies and laboratory chemical supply firms provide at least basic data for the products they manufacture or sell on their Web site. It is common to find material safety data sheets (MSDS) and technical bulletins online. For example, technical bulletins will often give extensive electrical properties for polymers used in electronics applications.

This is an obvious starting point for trademarked materials, but is also useful for basic, large volume chemicals. Once a manufacturer or supplier has been identified via standard sources, any good Internet search engine should readily provide a link to the company Web site. Chemical company URLs can be identified via Web directories such as:

- *ChemIndustry.Com* (http://www.chemindustry.com/)
- *MSDS Provider* (http://www.msdsprovider.com/Site/msdsprovider.nsf/search)
- *ChemExper Chemical Directory* (http://www.chemexper.com/)

The last two Web sites provide links to the supplier's Web site, at times directly to the full-text of the MSDS. *ChemExper* is discussed more fully in the Web site survey section of this article. Where the company name is known, use of a general search engine such as *Google* (www.google.com) can often identify the needed company home page just as quickly.

Four examples of the thousands of commerc
pages are:

- DuPont-Dow elastomers http://www.dupont-
- Occidental Chemical http://www.oxychem.c
- Acros-Fischer Companies http://www.acros
- Sigma-Aldrich Companies http://www.sigma

Most corporate sites require a simple registratio
uct literature. However, there is seldom a fee fc
triggering unwanted junk mail.

MSDS Compilations

There are several excellent Web guides and c
data sheets (MSDS). Keep in mind that MSD
supply firms, give only the basic properties, and
are left blank. Finding viscosity or thermal expan
unless that property is of particular importance i
use. A few of the best MSDS sites are listed in

SURVEY OF PHYSICAL PROPE

When searching for physical properties, one o
properties for common chemicals can be readil
covering physical properties: basic handbooks,
logs, and Internet sites. However, most patrons s
formation professional are more likely than
chemical or an unusual property or both.

The resources discussed in the rest of this pape
ing access to the major reference books and s
Landolt-Börnstein, STN International numerical

TABLE 1. Select MSDS Compilatio

NAME	# of MSDS
Cornell University	250,000
Vermont SIRI (mirror site)	180,000
Interactive Learning Paradigms Where to Find MSDS.	Web Guide
MSDS-Search	Web Guide

the government-sponsored thermodynamic data projects like Texas A&M University's TRC. Nor is there any substitute for a full retrospective search of the primary literature via *Gmelin*, *Beilstein*, and *Chemical Abstracts*. These points are clearly demonstrated by the comparison of coverage of the major free Web sites versus the subscription *CrossFire Beilstein* and *Gmelin* products available from MDL Information Systems. This comparison is presented after the discussion of the five largest property Web sites. However, there is a growing body of usable and generally reliable data available for free on the Web.

All the Web sites evaluated in this article are summarized in Tables 2, 4, 5, and 6. Column 2 (# of compounds) reflects the approximate number of compounds covered, which changes as the site is updated. Column 3 (max. # of properties) is the approximate maximum number of properties potentially available directly from the Web site. Some Web sites contain links to additional properties on other Web sites. The "Searchable by" columns reflect the ways in which the data can be searched. In order:

- Name–Chemical/common name
- CAS RN–Chemical Abstracts Service Registry Number
- MF–Molecular formula
- MW–Molecular weight
- Prop Values–Range-searchable property values
- Substruct.–Substructure searchable
- *ChemFinder*–If Y, Web site is covered by the *ChemFinder* metasearch site.

The Large Players: Recommended First Stops on the Web

Five Web sites have been identified that each cover more than 20,000 substances: *ChemFinder*, *NIST Chemistry WebBook*, *ChemExper Chemical Directory*, *Matweb*, and the *Physical Properties Database (PHYSPROP)* developed by the Syracuse Research Corp. The first two on this list are particularly important to consult in almost any physical property search.

ChemFinder (http://chemfinder.cambridgesoft.com/)

ChemFinder, the largest free property site on the Web, provides the structure, synonyms, CAS registry number, and up to nine physical properties directly for each compound (melting point (m.p.), boiling point (b.p.), refractive index, evaporation rate, flash point, density, vapor density, vapor pressure, and water solubility). The DOT number, EPA code, RTECS number, and a comment field with information such as the physical description and odor detection limits are also given, when available.

TABLE 2. Evaluated Web Sites: Large Pla

Web Site Name	# of compds	Max. # props	Searchable by: Name	CAS RN	MF	MW	Prop Values
The Large Players							
ChemExper Chemical Directory	70,000	4	x	x	x	x	x
ChemFinder	75,000	9	x	x	x	x	x
Matweb	25,412	75	x				x
NIST Chemistry WebBook	40,000	45	x	x	x	x	x
Physical Properties Database (PHYSPROP)	25,250	8		x			
General, Smaller Scale Web Sites							
Estimation Program Interface (EOI) Suite		14		x			
Hazardous Chemical Database	3,995	10	x	x	x		
Hazardous Substances Data Bank	4,500	30	x	x	x	x	
International Chem. Safety Cards	675	10	x	x			
NTP Chemical Health & Safety Data	2,000	28	x	x	x		
Organic Compounds Database	2,483	5	x		x	x	x

However, the best feature of *ChemFinder* is th gine, searching over 350 Web sites and displaying links are arranged into several categories inc MSDS, physical properties, regulations, structure fair number of the links are broken. A full listing o at http://chemfinder.cambridgesoft.com/about/che

The search engine works off a single master lis reviewed to eliminate obvious errors such as m CAS registry numbers. The search options are pa registry numbers, molecular formulas or weigh and substructures can all be used to locate right-hand truncation of chemical names is pro Three-dimensional structures can also be display life sciences and chemistry software solutions, c vices supplier, is to be commended for making this resource freely available, though one hopes that the bad links will be fixed. This free version limits retrieval sets to twenty-five compounds, is intended for occasional use, and comes at the price of numerous pop-up advertisements.

CambridgeSoft also offers annual subscription access for professionals and institutions to an enhanced version of the database under the name, *ChemINDEX*. *ChemINDEX* offers additional Web links, more generous retrieval limits, a data export feature, and no pop-up ads.

NIST Chemistry WebBook (http://webbook.nist.gov/chemistry/)

As befits the National Institute of Standards and Technology (formerly the National Bureau of Standards), the data quality and usability of this Web site is first rate. Up to forty-five thermochemical, thermophysical, and ion energetics critically reviewed properties are available for over 40,000 compounds. The data has been compiled from the NIST Standard Reference Data Program and outside contributors. The source of all data and, frequently, the method are carefully documented. Comments provide additional information such as the uncertainty of the measurement. Most of the data is in tabular form, but some can be displayed as X-Y plots. In addition, mass, IR, and UV/Visible spectra are provided in graphical form.

The database is predominantly organic with a few small inorganic compounds. As with *ChemFinder*, a full spectrum of search options exists including names, CAS registry numbers, property ranges, and substructure. Author searching permits one to check what literature references have been used in compiling the database. For each retrieved substance, basic information including molecular formula, synonyms, and structure are provided along with links to the various categories of properties.

ChemExper Chemical Directory (http://www.chemexper.com/)

This site provides a metasearch of over 70,000 chemicals from more than twenty supplier catalogs. The directory can be searched by registry number, molecular formula, chemical names, physical and chemical characteristics, and substructure. Links are provided to the supplier's Web site and to 16,000 MSDS, mostly from Acros. Only the basic properties are directly provided: density, m.p., b.p., and flash point. However, links to the full-text of the MSDS will usually provide some additional properties.

The ChemExper company, based in Belgium, encourages both commercial suppliers and academic laboratories to submit product information directly into the database. Though all organizations submitting data are registered, clearly the property data is not critically reviewe tion. However, this directory provides a quick v ties and laboratory-scale suppliers for a wide commercially.

Matweb (http://www.matweb.com/index.asp?ck

Although this article does not attempt to cov ized engineering materials resources, the *Matweb* sive, covering over 25,400 materials. Included superalloys, ceramics, glass, fibers, composites, Up to seventy-five properties are available for e done by material type, trade name, specificatio three different property ranges, and alloy compo ations Inc., designer and maintainer of this site, additional materials and data from companies. F ditional features and search options are availabl

Physical Properties Database–PHYSPROP (http://esc.syrres.com/interkow/PhysProp.htm)

The Syracuse Research Corporation has dev tal chemistry databases, supported partially by t tal Protection Agency (EPA), Procter and Gam contains chemical structures, names, and physi chemicals. Some of the values are estimated ra eight physical properties are provided: m.p., b.p ter partition coefficient, vapor pressure, pKa (a and OH rate constant in the atmosphere.

This free demo database can only be searche literature references are not given. Only the auth provided. Of course, with a little information de could be identified. The free database also d searching capabilities that are available with the Systems, Inc.) or the *Accord for Access* (Syno commercial versions of *PHYSPROP*. A fuller sources has been provided by Stoss.[6]

Comparison of the Large Free Sites vs. a Sub

A study was made of the number of proper sites described in the last section as compared t

sic *Beilstein Handbook* (organic chemicals) or *Gmelin Handbook* (inorganic chemicals). The *Matweb* site was excluded from this comparison since it deals mostly with materials, instead of individual chemical substances.

The *Beilstein* and *Gmelin* handbooks have been premier sources of physical property information for well over a century. Within the past decade, this information has been made available on the Web on a subscription basis by MDL Information Systems' *CrossFire* search system.

Ten chemicals were searched against *ChemFinder*, *NIST Chem WebBook*, *ChemExper*, *PHYSPROP*, and then *Beilstein* or *Gmelin* as appropriate. The chemicals were purposefully chosen to range from the very common (phenol) to a specialized research chemical *(1-(2-pyridinyl)piperazine)*. The results are given in Table 3. The number of references in CAS' *CAPlus* database shown in the third column provide a good indication of how common the substance is. Though limited in scope, this comparison shows the value and limits of free Web sources versus a high quality, but admittedly expensive, online database.

The number of unique properties, not including spectra, was carefully determined for each compound at each Web site. This data shows that the *NIST Chem WebBook* is the most comprehensive free Web site (184 total property values). The other three sites range from fifty-six values for *ChemFinder* to sixty-eight for *PHYSPROP*. However, if the data hidden in the 157 *ChemFinder* links were

TABLE 3. Comparison of the Free Web vs. a Subscription Service

Chemical	CAS Registry Number	CA Hits	*ChemFinder* (Directly Given)	*ChemFinder* (Metasearch links)	*NIST Chem WebBook*	*ChemExper* (incl. MSDS)	*PHYSPROP*	*Beilstein/Gmelin Crossfire*
Phenol	108-95-2	51,781	10	32	25	8	8	70
Aniline	62-53-3	34,059	9	26	28	9	8	68
Cyclohexane	110-82-7	25,539	8	25	35	8	7	72
Monochlorobenzene	108-90-7	14,255	9	29	26	10	7	69
Sodium Acetate	127-09-3	9,262	2	4	7	2	5	31
Phosphine	7803-51-2	6,275	4	17	15	0	7	43
Ciprofloxacin	93107-08-5 85721-33-1	6,155	0	1	0	0	5	12
Ethanethiol	75-08-1	4,152	7	12	26	10	8	47
Benzotrifluoride	98-08-8	1,023	5	9	22	7	7	40
1-(2-pyridinyl)piperazine	34803-66-2	545	2	2	0	5	6	7
Total Properties not counting spectra			**56**	**[Note 1]**	**184**	**59**	**68**	**459**

Note 1: A total of 157 sites are linked from *ChemFinder*, including the *NIST Chem WebBook* and *PHYSPROP* databases shown separately. Screening out overlapping and redundant properties was not attempted, given the large amount of time involved.

counted, it is expected that, in some cases, *ChemF*
ble the number of found properties. However, du
the additionally linked sites would be fairly high.

What really stands out are the properties avai
ucts, 459 properties which is 2.5 times more
WebBook. The free Web cannot match fee serv

General, Smaller Scale Web Sites

Estimation Program Interface (EOI) Suite
(http://www.epa.gov/oppt/exposure/docs/episuite.h

This handy software was developed by the S
der U.S. EPA contract. It is reviewed here beca
mation programs, it is available as freeware.
notation as the search key, results from ten sep

Aquatic toxicity (LD50, LC50)	Henry'
Aqueous hydrolysis rates	M.P, B
Atmospheric oxidation rates	Octanol
Bioconcentration factor (BCF)	Soil so
Biodegradation probability	Water

The program contains a SMILES notation da
istry numbers. By entering a registry number,
matically retrieved and entered into the search

Hazardous Chemical Database (http://ull.chem

The University of Akron maintains a databas
properties commonly found on MSDS for abo
there is little information that could not be foun
sent the data concisely and clearly.

Hazardous Substances Data Bank
(http://toxnet.nlm.nih.gov/cgi-bin/sis/htmlgen?

This database from the National Library of
anyone working in the environmental chemistry
tion to a very fine review of all environmental a

chemicals, the physical property section of these lengthy records provides up to ten of the more common properties. Unlike many free resources, the property values are critically reviewed and documented with the full literature reference.

International Chemical Safety Cards
(http://www.cdc.gov/niosh/ipcs/nicstart.html)

These safety cards are two-page summaries of basic hazard information for 675 of the most common chemicals in international commerce. The physical property section provides the typical eight to ten properties found on most MSDS. The World Health Organization, Commission of the European Communities, International Labour Organization, and NIOSH jointly sponsor this program.

NTP Chemical Health & Safety Data
(http://ntp-server.niehs.nih.gov/Main_Pages/Chem-HS.html)

The National Toxicology Program (NTP) was established in 1978 as part of the National Institute of Environmental Health Sciences (NIEHS). Its mission is to coordinate toxicological testing programs and disseminate information about potentially toxic chemicals. One result of the program is this Web site covering about 2,000 chemicals. The information is similar to the *Hazardous Substances Data Bank*. The typical properties for a database focusing on safety and handling are provided, such as flash point, evaporation rate, and pH. However, unusual properties, such as burning rate, are occasionally cited.

Organic Compounds Database
(http://www.colby.edu/chemistry/cmp/cmp.html)

Maintained at Colby College, this site features a database of 2,483 compounds compiled by Harry M. Bell of Virginia Tech. Though only a few common properties are provided, the search screen allows the selection of a wide variety of parameters including property values, element counts, and the presence or absence of certain broad structural entities such as amines or hydroxyl groups. Unfortunately, retrieval sets are limited to twenty compounds, though the search engine does report the total number of hit compounds.

Sites Focusing on Specific Types of Material

ARS Pesticide Properties Database (http://wizard.arsusda.gov/acsl/ppdb.html)

Developed by the U.S. Department of Agricu Service (ARS), this database is a compendium c erties of 334 widely used pesticides. Informatio cuses on sixteen of the most important propertie and degradation characteristics. References are values have been rechecked with the manufactu pesticide name only (see Table 4).

Critical Properties of Gases (http://www.flexwc

This simple, but useful table from Flexware pressure and critical temperatures for about 700 are both included.

Fuel Property Database (http://www.ott.doe.go

This database that provides key data on about pression ignition fuels, such as biodiesel and sy U.S. Department of Energy (DOE) Office of T cluded is information on various physical, chem

TABLE 4. Evaluated Web Sites: Spec

			Searchable by:				
Specific Types of Material Web Site Name	**# of compds**	**Max. # props**	Name	CAS RN	MF	MW	Prop Values
ARS Pesticide Properties Database	324	16	x				
Critical Properties of Gases	700	3	x		x	x	
Fuel Property Database	27	29	x				x
NIST Ceramics WebBook	265 families	40	x				
Plastics Additives Database	9,500	Varies	x		x		
Plastics Technology Materials Database	13,200 grades	Varies	x		x		x
Solv-DB	224	44	x	x	x	x	x

tal, safety, and health properties. The source and standard test methods used are also given.

NIST Ceramics WebBook
(http://www.ceramics.nist.gov/webbook/evaluate.htm)

The *NIST* (U.S. National Institute of Standards and Technology) *Ceramics WebBook* consists of two searchable databases, *High-Temperature Superconductors (WebHTS)* and *Structural Ceramics (WebSCD)*, and one browsable collection of property data summaries (*WebPDS*) for six categories of ceramics. As befits a NIST effort, the data source and quality are carefully reviewed and documented. Once a particular material has been retrieved, the format of the display is very similar to the better-known *NIST Chemistry WebBook.* Hyperlinked property categories take you to the exact point on the Web page displaying the information desired.

The Superconductor database provides evaluated thermal, mechanical, and superconducting properties for oxide superconductors. Materials are searchable by chemical family, informal name, structure type, and desired property. Authors' last name, publication source, and publication year are also searchable. The tables for a given property can be quite lengthy since they show values based on the mass fractions of the various component oxides.

The Structural database provides evaluated data for a wide range of structural, engineering, and fine ceramics. The search input form and display features are identical to the Superconductor database.

Plastics Additives Database
(http://www.specialchem.com/customers/demo/formSearchProduct_demo.asp)

Specialchem operates a set of free information services to link users and suppliers of plastic additives. A 9,500 product database from about 260 suppliers can be searched by additive function, base polymer, trade name, supplier, and keywords. The search can be limited to additives approved for food contact. Technical articles, datasheets, links to supplier Web sites, and Web forms for requesting quotations, technical data, or samples are available. Free registration is required to display much of the information.

Plastics Technology Materials Database
(http://www.plasticstechnology.com/dp/materials/)

Plastics Technology Magazine provides information on over 13,190 grades of plastics. They are searchable by about sixty properties, supplier name, resin family, descriptive terms like color, and price ranges. As expected, properties related to processing and use of plastics such as pact strength, and flammability are emphasized and easy to navigate. Particularly useful is the ab user-customized table format. What appears to b database is available from Bill Communications *terials Selection Database* (http://www.plaspec. that may be useful for discontinued products.

Solv-DB (http://solvdb.ncms.org/solvdb.htm)

Sponsored by the National Center for Manufa least 224 solvents can be searched by eight diffe vent name, CAS registry number, molecular for Nine different properties are range searchable inc sure, density, and surface tension. Up to thirty-thr played for each solvent. Results can be sorted by s range-searchable properties. Extensive informatio with display of health, safety, regulatory, and env

Sites Focusing on Specific Properties

Acoustic Material Property Tables (http://www.ultrasonic.com/tables/index.htm)

Specialty Information Associates maintains th tables in MS Excel format providing six differen rials. Solids, plastics, rubbers, liquids, and gases a gitudinal and shear piezoelectric information is

ATHAS Data Bank of Thermal Properties (http://web.utk.edu/~athas/databank/intro.html)

The Advanced Thermal Analysis Laboratory see and the Oak Ridge National Laboratory pro mal properties for 200 linear polymers and sma give values for temperatures from 0.1 to 1000° calculated values are clearly marked. Informatio and polymer melts.

Dielectric Constant Reference Guide (http://www.asiinstr.com/dc1.html#List)

Although this resource from ASI Instruments included because of the great variety of subst Where else can one readily find the dielectric co

TABLE 5. Evaluated Web Sites: Specific Properties

			Searchable by:								
Specific Properties Web Site Name	**# of compds**	**Max. # props**	Name	CAS RN	MF	MW	Prop Values	Substruct.	*ChemFinder*	**Type**	**Sponsoring Organization**
Acoustic Material Property Tables	470	6	x						Y	Table	Specialty Information Associates
ATHAS Data Bank of Thermal Properties	200	8	x						N	Table	Univ. of Tenn/Oak Ridge Natl Lab.
Dielectric Constant Reference Guide	1,500	1	x						Y	Table	ASI Instruments
Pesticide Fact Sheets–New Active Ingredients	42	16	x						Y	Table	U.S. EPA
Phase Diagrams Web	900		x						N	Table	George Tech Joint Student Chapter of ASM/TMS

flour? For materials with variable compositions such as polymers or minerals, a range of values is given. This is helpful in establishing the variation that might be expected in measuring a specific sample.

Pesticide Fact Sheets–New Active Ingredients
(http://www.epa.gov/opprd001/factsheets/)

This Web site contains extensive information on new pesticides registered with the U.S. EPA Office of Pesticide Programs. Since it covers only new active ingredients since fiscal year 1997, the file is small, only forty-two substances at this writing. However, as time goes on, it will grow in size and value. The fact sheets are listed by common name and can be over ten pages long. In addition to providing up to sixteen physical properties, the fact sheets provide use patterns, formulations, extensive toxicology information, and environmental fate.

Phase Diagrams Web
(http://cyberbuzz.gatech.edu/asm_tms/phase_diagrams/)

Maintained by the Georgia Institute of Technology Joint Student Chapter of ASM/TMS, most of the phase diagrams are for binary mixtures of elements, but there are some dealing with compounds and tertiary mixtures. A handy periodic table allows easy navigation of the listing. The site has a prominent disclaimer that provides no guarantee of accuracy. Although a hyperlinked source is given for each diagram, this link is often an individual's e-mail or is broken, even for those linking to an organization. Despite the source being specified on the Web page listing, no copyright statements accompany any of the diagrams. Caution in use of this data is recommended.

Academic Directories

The following university library Web pages a contain at least some links to public Web sites collection. However, even the indexes to their lc useful, since a searcher may well have access to s their location. Consider this listing to be represe ties maintain extensive Web directories in the sc

Arizona State University Index to Physical, Ch Property Data (http://www.asu.edu/lib/noble/c

This Web page starts with a brief, but well cl sources and then provides a list of over 400 prc print citations. The presentation of the informat numerous cross-references.

Duke University Chemical & Physical Propert (http://www.chem.duke.edu/~chemlib/properti

The Duke University properties page focu sixty-nine properties. However, the few Web li propriate.

TABLE 6. Evaluated Web Sites: Ac

			Searchable by:				
Academic Directories Web Site Name	**# of compds**	**Max. # props**	Name	CAS RN	MF	MW	Prop Values
Arizona State Univ. Index to Physical, Chemical, & Other Property Data		480					
Duke Univ. Chemical & Physical Properties in the Library		69					
Indiana Univ. CHEMINFO SIRCh Physical Properties							
Univ. at Buffalo Materials Properties Locator Database			x				
Univ. of Texas Thermodex		160	x				
Vanderbilt Univ. Finding Chemical & Physical Properties		120					

Indiana University CHEMINFO SIRCh Physical Properties (http://www.indiana.edu/~cheminfo/ca_ppi.html)

Any one spending any time in the field of chemical information quickly becomes familiar with this premier scientific information Web site maintained by Gary Wiggins of Indiana University. Though annotations are limited, the page is well organized. Many of the Web sites reviewed in this article are listed on this page. Of special value is the ability to keyword search across the entire *CHEMINFO* site and a back-of-the-book alphabetical index to the site.

University at Buffalo Materials Properties Locator Database (http://libweb.lib.buffalo.edu/sel/searchSelMaterials.html)

A searchable database of nearly 100 print sources is maintained by the Science and Engineering Library. The database record for each source contains a brief annotation, generic and specific property keywords, and types of material covered. An attempt was made to index every property covered in the source.

University of Texas Thermodex (http://thermodex.lib.utexas.edu/)

This also is a searchable directory focusing on thermodynamic properties and is maintained by the Mallet Chemistry Library. The search form also allows specification of classes of materials and specific compounds in combination with desired properties. A few publicly available Web sites are included in addition to the print sources.

Vanderbilt University Finding Chemical & Physical Properties (http://www.library.vanderbilt.edu/science/property.htm)

This is an extensive and easy-to-navigate Web directory arranged by over 100 physical properties. Free and subscription Web sources are included along with the print sources under each property. Unfortunately, there is no designation as to which Web links are subscription, i.e., available only to Vanderbilt patrons, and which are free.

CONCLUSIONS

The thirty Web sites reviewed in this article provide an impressive and useful array of physical property information. For finding common properties, the value of material safety data sheets and other company technical literature has been described. However, subscrip much as 2.5 times as many properties for a gi search of the free-of-charge Web sites. Physic Web is widely dispersed, requiring a patient an ous resources.

REFERENCES

Toxicology/Environmental

1. Barnes, Laura L. 2001. WebWatch: Enviror *nal* 126 (6):32-35.
2. Glander Hobel, Cornelia. 2001. Searching Internet. *Online Information Review* 25 (4): 257-66.
3. Greenberg, Gary N. 2002. Internet resourc mental health professionals. *Toxicology* 178 (3): 26
4. Riley, Ola Carter. 2002. Environmental hea our natural resources. *College and Research Librari*
5. Scarth, Lina Loos. 2002. Reference on the *Booklist* 98 (11): 963.
6. Stoss, Frederick W. 2001. The right chemis analysis and reference. *Internet Reference Services*

Biomolecules

7. Brzeski, Henry. 2002. An introduction to bic *lar Biology* 187: 193-208.
8. Sansom, Clare. 2000. Database searching w An introduction. *Briefings in Bioinformatics* 1 (1): 2
9. Weissig, Helge and Philip E. Bourne. 2002. *Crystallographica, Section D: Biological Crystallog*

Spectra

10. Avizonis, Daina and Shauna Farr-Jones. 20 netic resonance spectroscopists. *Methods in Enzymo onance of Biological Macromolecules, Part A*), 247
11. Taniguchi, Hisaaki. 2001. Databases and i trometry. *Saibo Kogaku* 20 (7): 1017-1027.
12. Youngen, Gregory Keith. 2000. A guide to lography. *Science & Technology Libraries* 19 (1): 4

Atomic Data

13. Dragoset, R. A. et al. 1998. NIST atomic and molecular databases on the World Wide Web. *NIST Special Publication* 926 (*Poster Papers-International Conference on Atomic and Molecular Data and Their Applications, 1997*), 24-27.

Property Calculation Software

14. Bachrach, Steven. 2002. *Chemistry software.* Cheminformatics Web site, Trinity University Dept. of Chemistry. http://hackberry.chem.trinity.edu/ChemistrySoftware.html.

15. Google Web Directory. 2002. *Chemical Engineering: Software.* http://directory.google.com/Top/Science/Technology/Chemical_Engineering/Software/.

Materials Science

16. Abrate S. 2002. World Wide Web resources for materials science. *Computer Applications in Engineering Education* 9 (4): 238-47.

17. Cebon, David and Michael F. Ashby. 2000. Information systems for material and process selection. *Advanced Materials & Processes* 157 (6): 44-48.

18. Pellack, Lorraine J. 2002. Materials science resources on the Web. *Issues in Science and Technology Librarianship* (34, Spring 2002). http://www.istl.org/02-spring/internet.html.

Printed Sources Reviews and Textbooks

19. Bottle, R. T. 1979. *Use of chemical literature.* Woburn, MA: Butterworth Publishers.

20. Buntrock, Robert E. 1992. Beilstein and Gmelin: classical chemical information for people who hate classics. *Database* 15: 104-6.

21. Maizell, Robert E. 1998. *How to find chemical information*, 3rd Ed. New York: John Wiley & Sons.

22. Mellon, M. G. 1982. *Chemical publications: Their nature and use*, 5th Ed. New York: McGraw Hill.

23. Wiggins, Gary. 1991. *Chemical information sources.* New York: McGraw Hill.

24. Woodburn, Henry H. 1974. *Using the chemical literature: A practical guide.* New York: Marcel Dekker.

Information-Seeking and Communication Behavior of Petroleum Geologists

Lura E. Joseph

SUMMARY. The petroleum industry has recently undergone considerable change due to the continuing conversion to a digital environment. These changes affect the way that petroleum geologists seek and communicate information in all areas of their work including the creation of prospects, well site work, development of petroleum fields, and keeping current with industry activity. The changes have affected all levels of communication, both within the company and between companies. Other changes in information-seeking and communication behavior are related to a greater emphasis on new technologies and unconventional resources and to various threats to geologic information. This paper provides an overview of the current infor-

Lura E. Joseph, MLIS, is Geology and Digital Projects Librarian and Assistant Professor of Library Administration, University of Illinois at Urbana-Champaign (E-mail: luraj@uiuc.edu).

The author would like to acknowledge and thank the following individuals for reading the manuscript and offering suggestions: Doug Johnson, Geologist at Chesapeake Natural Gas; Doug Strickland, Exploration Manager at Wolverine Gas and Oil Corporation; Gerald Meeks, Land Manager at Palmer Petroleum, Inc.; Tina Chrzastowski, Chemistry Librarian at the University of Illinois at Urbana-Champaign; and Hannes Leetaru, Petroleum Geologist with the Oil and Gas Section of the Illinois State Geological Survey.

[Haworth co-indexing entry note]: "Information-Seeking and Communication Behavior of Petroleum Geologists." Joseph, Lura E. Co-published simultaneously in *Science & Technology Libraries* (The Haworth Information Press, an imprint of The Haworth Press, Inc.) Vol. 21, No. 3/4, 2001, pp. 47-62; and: *Information and the Professional Scientist and Engineer* (ed: Virginia Baldwin, and Julie Hallmark) The Haworth Information Press, an imprint of The Haworth Press, Inc., 2001, pp. 47-62. Single or multiple copies of this article are available for a fee from The Haworth Document Delivery Service [1-800-HAWORTH, 9:00 a.m. - 5:00 p.m. (EST). E-mail address: docdelivery@haworthpress.com].

http://www.haworthpress.com/store/product.asp?sku=J122

10.1300/J122v21n03_04

mation-seeking and communication behavio dium-to-large petroleum companies in the America.

KEYWORDS. Petroleum industry, petro communication, digital, crude oil, natural

RECENT CHANGES IN INFORMATION

The petroleum industry has undergone consi five to seven years due to the continuing conver In 1991, Poland reviewed the literature on inform scientists and engineers, and correctly predicted technology would result in changes to informal 1991). Furthermore, there have been significa 1995 paper on the effects of technology on the i of scientists (Hallmark 1995).

Geologists formerly worked in a paper enviro books of production data, and paper logs. Mappi cil, and eraser, and final products were created o ment. Reports were typed by secretaries and te medium to large companies in the United State with digital data. The majority of a petroleum ge front of a computer workstation interpreting dat one to four monitors, side-by-side. Although som by hand, those results are usually given to a tech and layered with other information through the u Systems (GIS) software. GIS is a significant to data from diverse sources to create maps that a poses. For example, with the use of GIS soft powerlines, roads, buildings, lakes, wells, and p overlaid and compared in preparation for a seism

The use of the 3-D immersive visualization e sophisticated new technology that is becoming t dustry (Nelson, H. Roice Jr. 2000). The 3-D imm ogy incorporates geologic information includ faults, and well logs. Systems used in the petrole large-screen visualization systems, or virtual reality systems using head tracking and user interaction in real-time to increase immersion and the sense of reality (Midttun et al. 2000). The visualization centers enable a whole team to simultaneously view data in real-time. Norsk Hydro, Texaco, and ARCO built dedicated visualization centers in 1997, and many other companies have followed suit. According to Zeitlin, "3-D visualization capabilities . . . are rapidly moving into general application in most oil companies" (Zeitlin 2001). Zeitlin predicts that continuing advances in the technology will enable a greater amount of data to be processed with greater accuracy and in less time. Nonspecialists will be able to use the technology to understand and interpret data, and scientists will be able to move quickly from original reconnaissance of seismic data to interpretation. An example of a visualization center is Halliburton Company's *Magic Earth* <http://www.magicearth.com/default.asp>.

Information transfer has also changed. The petroleum industry has always been fast paced, with the most successful companies being those that have the most accurate, most current, and largest volume of relevant information. With the advent of the digital age the pace is faster than ever. Now most data, including well logs, production information, mud logs, and maps are transmitted via e-mail attachment, downloaded from the Internet, or sent on CD-ROM. There is less face-to-face interaction. Personal communication, both distance and intra-office, is commonly via e-mail or cell phone. A geologist in Oklahoma City can successfully monitor a well drilling in midwinter in British Columbia without ever visiting the actual site through the use of consultants, e-mail, and cell phones. If there is a need to view a sample or core, a digital photograph can be sent by e-mail attachment. Information can be transmitted and viewed on the computer screen or wireless system while being discussed via cell phone. The speed of transmission of data is in minutes or hours instead of days or weeks, and data and communication are shared by anyone, any place in the world, in real time. Many companies are switching to PowerPoint presentations in meetings. In 1980 Coppin and Palmer wrote about communication from the research laboratory to the operating company; no doubt the changes in technology have also changed communication in this area (Coppin and Palmer 1980).

There is a constant tension in communication of information between needing to keep company information proprietary for a time and needing unrestricted access to the information held by other companies. Exchange networks ("invisible colleges") of information are at least as important to a petroleum geologist as to a research geologist. The dispersal of petroleum company personnel from one company to others due to mergers and buyouts may actually have improved transfer of information. It is generally easier to acquire data and information from another company if an individual is able to deal directly with a friend who is now working for the other company.

Information storage is also being affected, wh and information-seeking behavior. Instead of pa information in digital files on servers. This appro dates to exhibits and reports. However, the use c rather than being reduced. Minor changes and have been overlooked due to the amount of time and typing. Now, multiple revisions may occur. on the company server, many geologists make p

Information-seeking is different today, but there is difficulty knowing where to look for companies, specific geologists or technicians a with new information vendors and sources and w ogy. There is a continual addition of new produ products. The learning curve is currently extren

Access to the journal literature is also cha have started publishing in electronic format and issues. Currently, American Association of P Geological Society of America (GSA), and Soc cists (SEG) are collaborating to create an ag Other societies such as American Geological In Society of London, Mineralogical Society of A mentary Geology (SEPM) have expressed inte erences in the journal aggregate will most likely other journal aggregates such as American Ge "any word" searching will be possible. Access journals will be included. Many of the smaller s tually be added. The ". . . goal is to have the ag braries in January 2004" (Noga 2002). Such certainly improve access to the literature for pe

Other recent changes in the information-seek ior of petroleum geologists are related to the fac petroleum in the contiguous United States has a geologists increasingly are searching for petrole or in deeply buried stratigraphic traps with no su cated technology, such as the use of 3-D seismic petroleum in complex and obscured areas. Alth data was formerly the primary responsibility of ward training petroleum geoscientists who can geological and geophysical data. Seismic data company or to a group of companies and may b pany or from vendors. Recent changes have ease

data (Strickland 2002). As the major petroleum companies have switched activities to international and Gulf Coast regions, they have begun to sell large portions of their geophysical data to brokers. It is now possible to purchase seismic data covering large areas of the United States for reasonable prices.

Due to the maturity of the petroleum industry in the USA, another change is the shift toward "unconventional" resources such as coal bed methane, tight gas sands, and shale gas (Freeman 2002). The search for unconventional resources broadens the types of information resources used by petroleum geologists to create prospects. In the case of coal bed methane, geologists seek information related to old coal fields and mines.

PETROLEUM GEOLOGISTS, ENGINEERS AND RESEARCHERS

Pinelli, and Ellis and Haugan have written about the differences in information seeking habits of engineers and research scientists (Pinelli 1991; Ellis and Haugan 1997). Petroleum geologists exhibit a blend of characteristics related to both groups. As with the engineer, the terminal degree for a petroleum geologist is generally a Masters Degree, whereas a research or academic geologist generally earns a Ph.D. While petroleum geologists may have values and methods of thinking closer to that of research scientists, the reality of working in a corporate environment may compel behavior more like that of an engineer. Typically the petroleum geologists' primary contribution is knowledge used to produce an end result; however, they also create new and original knowledge. The petroleum geologist's reward is monetary but also consists of recognition within the circle of the company and the profession. While some petroleum geologists may publish findings after the information is no longer proprietary, there is much less emphasis on publishing than for research/academic geologists.

TASKS AND RELATED INFORMATION BEHAVIORS

It is the goal of the petroleum industry to find accumulations of crude oil and natural gas in sufficient quantity and at shallow enough depths that it can be profitably produced and commercially developed. (For information on the subject of petroleum geology, consult any of a number of texts: Stonely 1995; Levorsen 1967; Selley 1997; Link 2001; Baker 1979.) The actual operation of finding and developing petroleum depends partly on the size of the company. Staffs of large and medium size corporations usually work in teams; a team may consist of a geologist, an engineer, a geophysicist, and a landman, each with separate tasks. In small companies one person may perform the tasks of several specialists, and one-person independents may d consultants for certain tasks. Even larger comp tasks. The type of information used and the co partly on the size of the operation. This paper is of the exploration or development geologist wo troleum company in the United States of Ameri

Task: Create Prospects

A large portion of the information-seeking a petroleum geologists is related to creating and p pect is essentially a proposal to drill a well, or w natural gas, and all the documentation necessa haviors relate to both the type of information u ing prospects.

Basic Information Resources Used by Petroleu to Create Prospects

There are a number of basic information re gists use to create prospects; these relate to both mation communication. These include various t production data, maps and cross sections, and

Several different types of well logs are ge wells. As the well is drilled the penetration rate penetration rate depends upon factors such as t the rock, and the type of bit being used. Penetrat nation with other information to determine form logs from previously drilled, nearby wells.

Sample logs and mud logs are other importa ing the drilling of a well, rock fragments (cuttin in the slurry of water and mud, and the solids an dard intervals, usually every ten feet of penetrat rial is collected in bags that are labeled with de depths these samples are collected by "rough helpers or floormen) (Baker 1979). When the such information is critical, consultants called mud logger will take over the job of collecting s the samples for rock type, porosity, and "show be recorded on a strip log in relation to depth of will also install equipment (a gas chromatogr ence, type, and amount of natural gas brought

mation is also displayed on the log. The shows of oil and/or natural gas indicated by the sample logs and mud logs are very important pieces of information used in deciding whether or not to complete a well. Even if the well is not completed, "shows" may help determine where to drill a subsequent well. Sometimes a different type of drill bit is used to obtain cores instead of cuttings. Coring a well is generally more expensive, but the resulting core can be analyzed to obtain more specific information about the formations being drilled. Sample logs, mud logs, and core information are generally proprietary information. Normally, this information is obtained by participating in drilling the well or by trading for the information. After a period of time, companies may send cores and samples to a storage facility and release proprietary rights.

Another essential form of information used by petroleum geologists consists of wireline logs. To log a well various measuring devices are lowered into the well on coaxial cable. As the instruments are drawn back up the well they continually measure properties of the rocks and transmit the information electronically to the surface where it is recorded. The resulting wireline logs indicate type of rock, porosity, and whether water, oil, and/or gas are contained in the pore spaces.

Most states require the release of certain information related to wells within a certain period of time. Often the quality of the information released and the time delay prevent this information from being of much use to the geologist; however, many companies release wireline logs to vendors within a reasonable amount of time. Wireline logs may be received as a result of participation in drilling wells or obtained from newly-drilled wells via personal contacts and from trades. There is always tension between wishing to restrict information, thereby maintaining the advantage of proprietary status, and the need to obtain other information that is being held by competitors. In some areas local societies maintain a library of wireline logs and other information such as sample logs, scout cards, and production books that their members may use. Previously, wireline logs were available on paper, but with the digital revolution, this information is now available in digital format. Examples of vendors include IHS Energy Group <http://www.pidwights.com/>, A2D <http://www.a2d.com/>, and MJ Systems <http://www.mjlogs.com/>.

Scout tickets or scout cards are another source of information often used by petroleum geologists. In the early years of petroleum exploration, "scouts" hired by companies would meet and trade information related to wells and activity. Now such information regarding completed wells is reported to vendors who compile the information in standard format and sell it back to companies. Information includes name and location of the well, company, total depth, tops of various formations, drill stem tests, shows, perforations, producing interval and initial flow rates (if applicable), and the final status of the well (oil producer, gas producer, dry and abandoned, plugge mation is now available in digital format from Group <http://www.pidwights.com/>. Canadiar AccuMap, Ltd. <http://www.ihsaccumap.com/> also be accessed on the Internet from some sta Department of Natural Resources <http://sonris-w sonris_portal_1.htm>.

Production information from completed well termine which formations are worth mapping a veloping into a field, for input into computer details, and for expert testimony at hearings. Prc production rates, are used by both geologists a stem tests may be conducted before a well is com is likely to be commercially productive. Before t lic knowledge, geologists and engineers seek it t as scouts and pumpers and by participating in w nies. Some production information is now ava commercial vendors, and from state agencies vi the former is PI/Dwights PLUS Production Data products/information/pidwightsplusproductiond are North Dakota Industrial Commission Oi explorer.ndic.state.nd.us/>, Wyoming Oil and C <http://wogcc.state.wy.us/wogcce.cfm>, New Y vironmental Conservation <http://www.dec.sta htm>, and Louisiana Department of Natural R dnr.state.la.us/www_root/sonris_portal_1.htm> tion such as rate, pressure, and choke size can n to the operating company via satellite.

Maps and cross sections are major tools used information. They are created as part of a pro maps and cross sections are also used to gain Available regional maps and cross sections are logic context. Such resources may be produce geologic surveys, or they may already exist as p Other maps and cross sections may be acqui through examining prospects from other compa operated by partners, or from previous work wi

Knowledge of the legal spacing requirement formations is also essential to both landmen an field rules govern the number of wells that may b of this information is now available on the Inter sites of the Texas Railroad Commission <http://ecap.rrc.state.tx.us/Apps/WebObjects/DrillingPermits.woa/wa/Query/fieldQuery> and the Louisiana Department of Natural Resources <http://sonris-www.dnr.state.la.us/www_root/sonris_portal_1.htm>.

Important information is sought and found in the memories and experience of fellow geologists and engineers, and it is common to consult experienced colleagues both in the company and in the industry. This approach is especially helpful when beginning work in an unfamiliar area. Colleagues can quickly provide a background on the regional geology, the producing formations, and the accepted interpretation of the area, and can suggest literature and other information resources. Much of this type of knowledge is being lost due to acceptance of early retirement offers and layoffs related to mergers and buy-outs.

The Process of Creating a Prospect

Specific information-seeking and communication behaviors are related to each phase of prospect creation. Every prospect begins with an idea. As one seasoned petroleum geologist put it "If I were a drop of oil or an mcf of gas, where would I hide?" As Pratt phrased it, "Where oil is first found, in the final analysis, is in the minds of men" (Pratt 1952). An idea may be sparked in any number of ways. Geologists are often assigned to geographic areas. When first beginning work in a new area, a geologist may create regional maps to gain an understanding of the area. Most domestic areas have already been drilled, but with a fresh approach the new regional maps generally trigger ideas for new prospects. Examination of well logs and scout card information, whether old or new, may reveal new or overlooked possibilities. Companies often lower the risk of drilling an initial well by seeking partners, and information gained from inspecting prospects from other companies may trigger new ideas. Participating in a well drilled by another company also allows early inspection of proprietary information during and immediately after drilling, and this may lead to a new prospect. Ideas may be spawned while prowling through old well files owned by the company or gained in mergers. Various petroleum news periodicals are available and may stimulate ideas for new prospects. These will be further covered in the section on current awareness. Information may also be gained by driving the county roads and looking for gas flares from drilling wells. Other activities include counting stands of pipe to determine depth of drilling wells, and sitting in cafes in rural areas near drilling activity and listening to conversations. Nearly any piece of information encountered while seeking or communicating other petroleum related information may lead to a prospect if one is creative.

Literature searches are probably less common in the petroleum industry than in academia, given their different objectives. Generally, the petroleum geologist will not be publishing a paper, so there w search. However, especially when beginning wc may conduct a literature search of its geology. A used to determine "analog fields" for compariso

GeoRef is probably the database most commo ture searches in the United States. According American Geological Institute (AGI), most of th have access to GeoRef in at least one location, an tion offices will use online services such as Dial Tahirkheli sees a trend toward obtaining a world in order to make the database available throug constant mergers and closures of companies hav tions, Tahirkheli suspects that the total numbe creased with desktop access; the total number of unknown to AGI, however, due to the proprietar

If a company does not have access to indexes su may travel to a local university to conduct literatu resources; alternatively, an information consultan hired to search and provide document delivery. M have small libraries with literature resources. Lar search or information facility with several informa sources that can be provided to offices througho

Geologic field trip guidebooks are good sourc geology. Other information sources include state and maps, society publications, dissertations Bichteler has discussed the difficulties encountere sorts of information, sometimes called "gray liter background information can be found in journals,

Nearly every prospect will include a productio and type of wells (gas, oil, dry holes, plugged wel of the wells, producing formations, "shows," am perhaps, projected ultimate production. The infor from commercial scout cards, commercial produ ords. Other current information not available com personal contacts and partners. The "base map" (m wells) may be commercially produced and will be other maps. Digital data are now available for the

After creating a production map, the next step cross sections from well logs which are "hung" fro A stratigraphic cross section is hung from a partic tion top or base. Subsequently, the tops of importa on each log, and correlated from well to well. All other wells in the area will then be correlated to the cross section wells, and important tops of formations will be noted. A structural cross section is normally related to sea level, and also includes interpreted folds and faults. Other information such as gas/oil/water contacts, petroleum shows, tests, and producing intervals may be communicated via the cross sections. Cross sections were formerly created by hand using logs in print format. Now they are often created from digital logs using computer software.

The formation top data from the correlated logs are used to create various maps that illustrate why a well should be drilled. These maps may include structure maps, and various isopach maps that indicate thicknesses of formations and rock units. Maps are now created at computer workstations using digital data and computer software. Most prospects include an analog field. The geologist is essentially saying "We expect our prospect to be similar to this field which has been commercially successful." The analog field will be illustrated with its own set of cross sections, maps, and production information.

Although a large portion of communication is visual, a prospect will also include a short report summarizing the information included on the maps and cross sections, and predicting the amount of "pay" (thickness of the producing interval) and the amount of petroleum that is expected to be recovered. The report will include a statement of comparison with the analog field. A statement of risk may also be included. (Some companies run a computer model to indicate the potential risk of a drilling venture.)

Various meetings related to prospects are held. Most medium-to-large size petroleum companies have an annual budget meeting at which time prospects for the coming fiscal year are presented. One person, or an entire team, may present each prospect to management. Throughout the year there are normally departmental and/or team meetings to discuss the status of various prospects. These meetings are generally much less formal than annual budget meetings, however, the mode of communication is generally the same: visual aides consisting of maps and cross sections, and oral presentation of reports. Others on the team may present geophysical, engineering, and lease information. During or after these meetings, maps, cross sections, and reports may be revised, and these revisions become part of the company information record. Before the meetings, team members generally communicate face-to-face, by phone, and by e-mail to exchange information and plan for the presentation. Sometimes an elaborate "montage" is created as a visual aide. The montage will incorporate all the maps, cross sections, and possibly geophysical lines and an abbreviated report. Previously, the geologist worked with pencil and eraser on paper, and the drafting department produced the final product. Increasingly, the geologist creates the maps and cross sections at a computer workstation. A technical department may produce the final product.

Many petroleum companies prefer not to retai prospect or well. They may dilute the economic hole (nonproducing well) by finding partners. In c gist or team will present the prospect to interested information will be shared in order to protect as n possible. Nevertheless, other companies are able information by reviewing prospects. Companies with whom they pair regularly due to a good past general information is presented, including cross s production information and analog field example

Petroleum-producing states have agencies that petroleum production. States may require a heari The regulating body varies from state to state, as dures. Generally the same set of maps, cross-sec hearings, and the testimony may include a geolog ten, any party with an interest in the subject acrea hibits. This is one way to obtain information not p If a party is protesting the drilling of the well, both a team of experts to testify, and the whole process which may be sought, communicated, and filed.

After a prospect has been drilled, generally a sections, reports, and other documentation are pl through old prospect files can be a worthwhile ac never drilled or shows may not be pursued. Ec change. If a dry hole has been drilled, it is often n immediately drill a subsequent well nearby. A pe to resurrect a prospect, or create a new one fron files. With the advent of digital files informatio ever, many geologists still keep paper back-up f

Task: Well Site Work

Another area of responsibility for the petrole work. Over the past twenty years, geologists wc perienced less actual on-site well duty. Consult this work with the geologist remaining in contac ports. Once a mud logging unit is on site, the geo report from the mud logger, and this information members of the team, and partners. Informatio well, penetration rate, likely formation tops, "sh difficulties encountered such as stuck drill pipe

drogen sulfide gas. Formerly, a copy of the current mud log was mailed or faxed to the geologist. Now, the current mud log is often sent as an e-mail attachment in LAS (log ASCII standard) format, although the fax machine is also still used. Some mud logging companies store the mud log on a server, and the geologist can access it via the Internet. A geologist and/or engineer may still travel to the well site to look at the drilling logs and mud logs and to consult with the mud loggers, especially when a well is being logged or cored.

Task: Help Develop Oil and Gas Fields After Discovery

In some petroleum companies there are separate exploration and development departments. The exploration department is in charge of drilling the initial well, and if successful, the development department is responsible for subsequently developing the field. In other companies, one department is responsible for both functions. In either case, engineers generally take more of a leading role in the development process. The same types of information are used, and the same communication behaviors apply, although a larger portion of the information will likely be owned by the company.

Task: Keep Up with Industry Activity

It is imperative for petroleum geologists to keep up with industry activity, including current knowledge of new finds and new areas being worked. Some of the resources for current awareness include trade journals such as *Oil and Gas Journal* and *Hart's E&P*, and reports such as *PI/Dwight's PLUS Energy News on Demand*, available as regional coverage by subscription from IHS Energy Group <http://www.pidwights.com/products/information/enod/>. Another resource is *Petroleum Abstracts Bulletin*, which is now sent weekly by e-mail, and consists of abstracts of journal articles, technical papers, and patents covering the worldwide petroleum exploration and production industry <http://www.pa.utulsa.edu/PA/pet_abs.html>. Petroleum geologists also scan new scout card information and new well logs. Other current awareness information is gained from "gray literature," conversations with engineers and rig personnel, lunches with contractors, and participation in drilling of wells. Information about new techniques can be obtained by attending conferences, field trips and field courses, workshops, and continuing education courses.

THREATS TO INFORMATION

There are several current threats to geological information. One such threat is created by mergers and buyouts of petroleum companies. When companies merge or are bought, some of the personnel are redundant. In many cases, experienced geologists leave the profession altoge ing another job, disillusionment, or new op experienced personnel leave the profession, a g This may happen to a lesser extent when a perso time and incentive to pass along information. A are bought out, files are sometimes purged. The rapidly and not necessarily with care. Some we cates, when in reality a file may contain extra i tions on logs, scraps of paper with notes, and r anywhere else. The significance of some prospe stood, and therefore discarded. When this info forever. Peripheral information, such as inform logs by geologists, can also be lost when older

Another threat to information was the focu *Geotimes* (Cutler and Maples 2002; Keane 20 2002; Tinker 2002). The core and sample data c peril. According to Tinker,

> The U.S. petroleum industry has reached the great private petroleum research labs the 1900s have been closed. And the data core and cuttings databases once housed considered just annual costs by many c them. (Tinker 2002)

Some collections are in danger of being aban the high expense of maintenance. Other collectio quately. In some cases the existence of the collec quate curation and the lack of sufficient or acce Research Council established the Committee on Data and Collections and recently released a rep ommended building three regional geoscience d proposed criteria to assess which data collection centers and adequate cataloging and metadata w formation and help petroleum geologists locate p

CONCLUSION AND PRE

The opportunities offered by the new digital to sweeping changes in the petroleum industry of information, but also the ways that petroleum cate information in all areas of their work. These changes have affected all levels of communication, including intra-company and inter-company exchanges. More and more information will likely be provided in digital format. Discovering what information is available, especially via the Internet, is currently frustrating and will likely continue to be a problem. Library professionals can offer a much needed service by ferreting out sources of information such as production data, field rules, well logs, and scout information, and creating Web pointers to collections of data.

Other changes in information seeking and communication behavior are related to a switch in focus to new technologies and unconventional resources. As the more easily found petroleum is depleted, ever more emphasis will be placed on developing new technologies for use in areas of complex geology, and for unconventional resources. The learning curve will likely remain steep. New technologies such as GIS and visualization centers, as well as others not yet imagined, will continue to be developed and will change the profession. The borders between professions will continue to blur as the tasks of geologists, geophysicists, and computer specialists blend.

The various threats to geologic information such as company mergers and loss of stored samples and cores could have a serious negative impact on geologic information unless current trends are reversed. Some experienced geologists believe there is a danger that geologists new to the profession will rely solely on the digital technologies and thereby be too removed from reality. They believe that something is lost in never looking at the actual rocks in situ and never hand-contouring maps or visiting drilling wells. Nevertheless, the advantages offered by the digital revolution greatly outweigh any negative aspects.

REFERENCES

Baker, Ron. 1979. *A Primer of oilwell drilling*. 4th ed. Austin, TX: The University of Texas at Austin.

Bichteler, Julie. 1991. Geologists and gray literature: Access, use, and problems. *Science & Technology Libraries*. 11(3):39-50.

Coppin, A.S., and Palmer, L.L. 1980. From the research laboratory to the operating company: How information travels. *Special Libraries*. 71(7):303-309.

Cutler, Paul, and Maples, Christopher G. 2002. Resources in peril. *Geotimes*. 47(6):16-19.

Ellis, David, and Haugan, Merete. 1997. Modeling the information seeking patterns of engineers and research scientists in an industrial environment. *Journal of Documentation*. 53(4):384-403.

Freeman, Diane. 2002. Rocky basins assessment due; Partial results set for Laramie meeting. *AAPG Explorer*. 22(9):37.

Hallmark, Julie. 1995. The effects of technology on t of Scientists. *Proceedings of the 29th Meeting of t ety, Seattle, Washington*, p. 51-56.

Keane, Christopher M. 2002. Data preservation: A 47(6):20-21, 45.

Levorsen, A. I. 1967. *Geology of Petroleum*. San Fran pany: 724 p.

Link, Peter K. 2001. *Basic Petroleum Geology*. 3rd. e tants International.

Midttun, Mons, Helland, Rolf, and Finnstrom, Erik value to exploration and production. *The Leading*

Mikulic, Donald G., and Kluessendorf, Joanne. 2002 tions. *Geotimes*. 47(6):24-26.

Nelson, H. Roice Jr. 2000. Immersive visualization. *Th*

Noga, Michael M. 2002. President's Column. *Geosc letter*. 195:1,3.

Pinelli, Thomas E. 1991. The information-seeking h *Science & Technology Libraries*. 11(3):5-25.

Poland, Jean. 1991. Informal communication amon view of the literature. *Science & Technology Libr*

Pratt, Wallace E. 1952. Toward a philosophy of oil- *Association of Petroleum Geologists*. 36(12):223

Selley, Richard C. 1997. *Elements of Petroleum Ge* demic Press, 496 p.

Stoneley, Robert. 1995. *An Introduction to Petroleum* New York: Oxford University Press, 199 p.

Strickland, Doug. 2002. Communication, 3 Septemb

Tahirkheli, Sharon. 2002. Communication, 24 July.

Tinker, Scott W. 2002.Valuing Earth's books. *Geoti*

Zeitlin, Michael. 2001. How 3-D visualization will c future oil companies. *The Leading Edge*. 20(12):

Online Bibliographic Sources in Hydrology

Emily C. Wild
W. Michael Havener

SUMMARY. Traditional commercial bibliographic databases and indexes provide some access to hydrology materials produced by the government; however, these sources do not provide comprehensive coverage of relevant hydrologic publications. This paper discusses bibliographic information available from the federal government and state geological surveys, water resources agencies, and depositories. In addition to information in these databases, the paper describes the scope, styles of citing, subject terminology, and the ways these information sources are currently being searched, formally and informally, by hydrologists. Information available from the federal and state agencies and from the state depositories might be missed by limiting searches to commercially distributed databases. *[Article copies available for a fee from The Haworth Document Delivery Service: 1-800-HAWORTH. E-mail address: <docdelivery@haworthpress.com> Website: <http://www.HaworthPress.com> © 2001 by The Haworth Press, Inc. All rights reserved.]*

Emily C. Wild is affiliated with the U.S. Geological Survey, 275 Promenade Street, Suite 150, Providence, RI 02908 (E-mail: ecwild@usgs.gov). W. Michael Havener is affiliated with the University of Rhode Island, Graduate School of Library and Information Studies, Rodman Hall, Kingston, RI 02881 (E-mail: mhavener@uri.edu).

The authors would like to thank the following individuals for their helpful comments on earlier drafts of this paper: Paul Barlow, Jim Campbell, Mike Eberle, C. Lee Regan, Angel Martin, Jr., and Gail Moede, U.S. Geological Survey.

[Haworth co-indexing entry note]: "Online Bibliographic Sources in Hydrology." Wild, Emily C., and W. Michael Havener. Co-published simultaneously in *Science & Technology Libraries* (The Haworth Information Press, an imprint of The Haworth Press, Inc.) Vol. 21, No. 3/4, 2001, pp. 63-86; and: *Information and the Professional Scientist and Engineer* (ed: Virginia Baldwin, and Julie Hallmark) The Haworth Information Press, an imprint of The Haworth Press, Inc., 2001, pp. 63-86. Single or multiple copies of this article are available for a fee from The Haworth Document Delivery Service [1-800-HAWORTH, 9:00 a.m. - 5:00 p.m. (EST). E-mail address: docdelivery@haworthpress.com].

http://www.haworthpress.com/store/product.asp?sku=J122

10.1300/J122v21n03_05

KEYWORDS. Hydrology, U.S. Geologica online databases, online indexes, ground sources

INTRODUCTION

Hydrology is the study of the distribution and teractions with land. Surface-water hydrologists streams, lakes, and ponds; ground-water hydrolc surface environments. Federal, state, and local a private sector all employ hydrologists. In an inte drology, professionals may have academic backg neering, environmental engineering, chemistry, hydrology are often interdisciplinary and are pub cooperative efforts between universities and fede agencies, and state agencies and private consultar terials for a literature review about the dynamics quire searching databases in all of the interdisci ogy, and the identified materials may include doc and private sectors. This paper provides an unde and other geoscientists obtain bibliographic in they use.

Scientific researchers often find publications other than traditional bibliographic sources like bases maintained by commercial and non-comm tives include referrals from colleagues and publication sources available from federal and U.S. Geological Survey (USGS) and state geolo of this paper are to (1) describe online sources of publications in hydrology; (2) explain the disser the USGS Water Resources Discipline, one hydrologic data and information in the United St of terminology and syntax that will improve se this paper lists selected online sources for citatio gies that were used in compiling a bibliography f intrusion along the Atlantic coastal zone of the U is part of the USGS Ground-Water Resources F program assesses the movement of the saltwate and upward toward public-water supply withdra fers. In 1995, it was estimated that over 30 mill

coast were dependant on freshwater aquifers for drinking water (Barlow and Wild, 2002). The encroachment of saltwater can result in the abandonment of public-supply wells because the water quality does not meet drinking-water standards (Barlow and Wild, 2002).

The bibliography for the saltwater-intrusion assessment was prepared largely from online bibliographies, publication lists, and library catalogs available from USGS offices and state geological surveys along the Atlantic coast. In addition, commercial and non-commercial bibliographic databases were queried to expand on the core list of citations and thereby ensure a comprehensive bibliography of publications. For comparison purposes, and to assess differing search strategies, searches in commercial and non-commercial databases were also done, generating citations in a variety of formats.

FIGURE 1. Selected Areas Along the Atlantic Coast Where Saltwater Has Intruded Freshwater Aquifers (from Barlow and Wild, 2002)

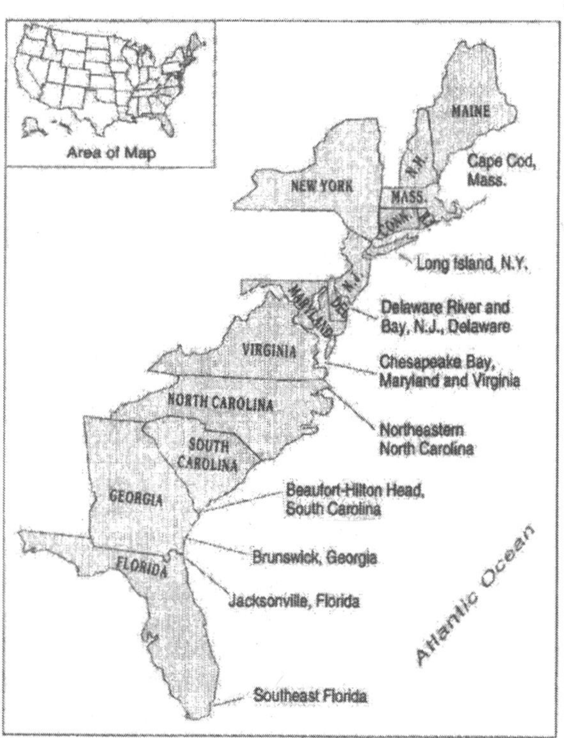

IMPORTANCE OF ALTERNAT *OF INFORMATION IN HY*

Hydrologists tend to do literature searches th use of traditional bibliographic indexes and data as in other disciplines, commonly starts with cons the specialty. One cannot ignore the importance o tions between hydrologists, either at the beginni drological investigation, as an information-gat supplemented with more traditional bibliograph common practice for hydrologists within the USC who are doing or have done similar research stud ographies of their colleagues' publications. Fu means of bibliographies provided through online ever, in some cases, hydrologists do not use bibl are not aware that such sources are available.

Often, time is a major constraint that can det use of bibliographic databases. In the public and reviewing available literature before starting pro port research may be among the most important to ensure a high-quality product, but this is often which the least time is allocated. Therefore, hy tend to depend on informal colleague-to-collea written and oral recommendations of publica within hydrology. For example, a hydrologist be trusion would contact colleagues who have previ lyzed freshwater and saltwater interactions in g hydrologist would collect bibliographic inform lection of publications. A hydrologist's use of fe ographies and publication lists can provide al information to that obtained from personal cont

Bichteler and Ward (1989) studied the geoscientists. Those interviewed included ge private, and government sectors. The individu time was the main factor that affected the way i tion information. Some individuals stressed the using professional contacts for bibliographic i on the traditional means for citation retrievals geoscientists emphasized that time, physical ac graphic control of gray literature (documents o publication distribution) were the principal problems with the use of indexes and databases as tools for gathering information.

The limited access to citations to gray literature produced by government agencies is a particular problem (Bichteler, 1991). For example, many hydrologists depend on thematic maps, such as those depicting ground water, bedrock geology, and surficial geology, to assess an area of study for hydrologic research. Citations to thematic maps would not necessarily be accessible through traditional bibliographic databases, but they would be accessible through direct contact with the USGS or state agency in the geographic area of interest. Often, gray literature is indexed as a monograph or a series (if indexed at all), and the limited print runs of these materials can restrict the quantity of publications disseminated. With the capability of the Internet to disseminate federal and state bibliographic sources of information, however, access to gray literature is improving. For example, many USGS state offices provide full-text documents, such as PDF files, through links or online publication lists.

From the geoscientists' perspective, direct communication with colleagues to obtain bibliographic data seems more efficient than use of traditional bibliographic sources, although effective communication between librarians and geoscientists may improve access to gray literature and facilitate geoscientists' ease in using traditional library reference tools. Nevertheless, for the immediate future, the combination of traditional and nontraditional bibliographic sources provides the best approach for geoscientists–including hydrologists–to do an inclusive bibliographic search for topical information.

U.S. GEOLOGICAL SURVEY SOURCES

Overview of the U.S. Geological Survey

The USGS is one of the largest earth-science organizations in the United States, providing publications and other earth-science information to geoscientists and the public. The mission of the USGS is to provide "reliable scientific information to describe and understand the Earth; minimize loss of life and property from natural disasters; manage water, biological, energy, and mineral resources; and enhance and protect our quality of life" (U.S. Geological Survey, 2000a). The USGS has four scientific disciplines (Biological Resources, Geological, National Mapping, and Water Resources). The Biological Resources Discipline (BRD) has seventeen research centers within the eastern, central, and western regions of the United States (http://biology.usgs.gov). The Geological Discipline (GD) comprises science teams performing research in the eastern, central, and western regions (http://geology.usgs.gov). Five regional mapping centers constitute the National Mapping Discipline (NMD) (http://mapping.usgs.gov).

The Water Resources Discipline (WRD) ach of the USGS by providing information on the us in the United States through professional collab agencies at the local, state, and federal levels (U The WRD has offices in each state, as well as in the WRD has offices relating to various sub-dis tional Headquarters in Reston, Virginia. These Quality (http://water.usgs.gov/owq), the Office usgs.gov/osw), and the Office of Ground Wat The focus of this paper is primarily on hydrolo through the WRD and the WRD's state offices. can be accessed at http://sa.water.usgs.gov (Tabl the postal state abbreviation (for example, fo http://ma.water.usgs.gov for information with re

Distribution Structure of U.S. Geological Sur

Because the USGS is involved in hydrologic of government, it is appropriate to assess the av hydrologic publications at the local, state, regic levels (Figure 2). Often, dissemination of hydr

TABLE 1a. U.S. Geological Survey Water Re Along the Atlantic Coastal Zone–Bibliographi Level Web Addresses Are Subject to Change)

Source–Web address	Publication page-Web address-c
Connecticut *http://ct.water.usgs.gov*	*Selected Reports on Geology and* *http://ct.water.usgs.gov/index/ctbib*
Florida *http://fl.water.usgs.gov*	*Water Resources of Florida Bibliog* *http://fl.water.usgs.gov/Pubs_produ*
	Florida Saltwater Encroachment Bi *http://fl.water.usgs.gov/Pubs_produ*
Georgia *http://ga.water.usgs.gov*	*Recent Publications–Georgia biblio* *http//ga.water.usgs.gov/ga004.htm*
	Georgia Fact Sheets Online–Links *http://ga.water.usgs.gov/publicatio*
Maine *http://me.water.usgs.gov*	*Maine bibliography published or re* *http://me.water.usgs.gov/mebiblio.*
	Recently published reports (full-tex *http://me.water.usgs.gov/newrepor*
Maryland-Delaware-District of Columbia *http://md.water.usgs.gov* *http://de.water.usgs.gov* *http://dc.water.usgs.gov*	*Online Reports–http://md.water.us* *Bibliography for MD-DE-DC (1886-* *http://md.water.usgs.gov/publicatio*

TABLE 1b. U.S. Geological Survey Water Resources Division State Offices Along the Atlantic Coastal Zone–Bibliographic Information Sources (Lower Level Web Addresses Are Subject to Change)

Source–Web address	Publication page-Web address-characteristics
Massachusetts-Rhode Island *http://ma.water.usgs.gov* *http://ri.water.usgs.gov*	*Reports Published or Released by District* 1977-1998 *http://ma.water.usgs.gov/USGS_MA_Biblio.htm* 1999-2001 *http://ma.water.usgs.gov/pub_99_01.htm* *Rhode Island Subdistrict Publications* *http://ma.water.usgs.gov/pub_RI.htm*
New Hampshire-Vermont *http://nh.water.usgs.gov*	*Water Resources of New Hampshire and Vermont, Bibliography of New Hampshire/Vermont District 1986-2001* *http://nj.water.usgs.gov/Publications/bibliography01_subjectWeb.htm*
New Jersey *http://nj.water.usgs.gov*	*Water Resources of New Jersey–Bibliographic Search Page* *http://nj.water.usgs.gov/pub/bibsearch.html* *Publications Available in Electronic Form* *http://nj.water.usgs.gov/pub/onlinepubs.html*
New York *http://ny.water.usgs.gov*	*Online New York State Reports, Fact Sheets, Articles and Abstracts* *http://ny.water.usgs.gov/htmls/pub/publist.html* *New York District Publications Search Engine (searchable index via author, titles, keywords)* *http://ny.water.usgs.gov/pub/bibsearch.html*

TABLE 1c. U.S. Geological Survey Water Resources Division State Offices Along the Atlantic Coastal Zone–Bibliographic Information Sources (Lower Level Web Addresses Are Subject to Change)

Source–Web address	Publication page-Web address-characteristics
North Carolina *http://nc.water.usgs.gov*	*North Carolina Online Bibliography* (searchable bibliography) *http://nc.water.usgs.gov/pubs/index.html* *North Carolina Online Bibliography–Publications Retrieval for Online Reports* *http://sun3dncrlg.er.usgs.gov/cgi-bin/pubs?keyword=Full+Text*
Pennsylvania *http://pa.water.usgs.gov*	*USGS Pennsylvania Publications* (online publications and searchable bibliography links) *http://pa.water.usgs.gov/pa_pubs.html*
South Carolina *http://sc.water.usgs.gov*	*Water Resources of South Carolina publications* (some online) *http://sc.water.usgs.gov/publications/recent_pubs.html*
Virginia *http://va.water.usgs.gov*	*Water Resources of Virginia Publications* (Recent Publications, On-Line Reports, and Bibliography) *http://va.water.usgs.gov/publications.html* *Chesapeake Bay References–http://chesapeake.usgs.gov/references.html*

graphic sources pertaining to a specific topic or cal level within a region. Over time and throug distribution processes, localized publications (a seminated to a wider audience over increasin graphic sources. As the bibliographic informati becomes available at higher levels of distributio citations that are indexed becomes more variab commercial bibliographic sources essentially rather than comprehensive collections of all rel

Understanding variations in the disseminati maps, and other formal and informal series can cess to USGS bibliographic information. An im all government documents are published and ind Printing Office (GPO), as indicated within so sources. It is commonly stated in some library s and librarians can use GPO indexes to obtain ever, locally published and indexed government in the GPO sources. In the USGS publication se ries (described in the next paragraph) are publish through services provided by GPO (report glob formal reports are printed through arrangements ment thereafter is reduced; these reports are dist a local or regional level through USGS state o

FIGURE 2. Indexing and Availability Trends of U. tions in Hydrology

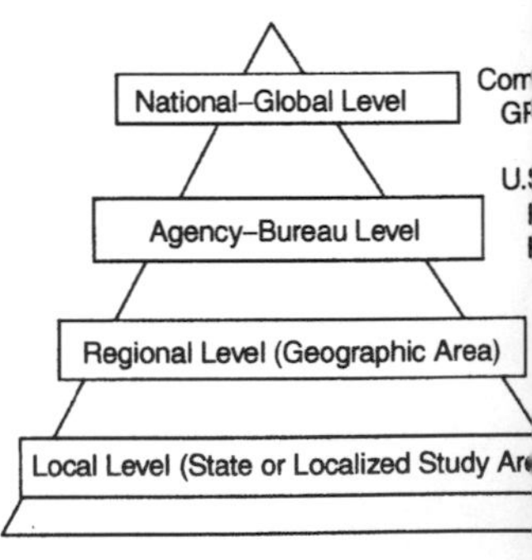

Therefore, informal reports are not likely to be included in the major commercial databases.

Materials authored by USGS scientists include books, information periodicals, informal reports, thematic maps and charts, outside publications (predominantly journal articles and conference papers), *Fact Sheets*, and topographic maps (U.S. Geological Survey, 2000c). Each publication series has a distinctive technical level and target audience. Sources for bibliographic information of formal and informal reports are available online through USGS information sources (Tables 2a, 2b, and 2c). Selected publications are also available online, and are posted by the USGS GD (Tables 2a, 2b, and 2c). Some examples of the formal series published by the USGS are *Professional Papers*, *Bulletins*, *Water-Supply Papers*, *Circulars*, and *Techniques of Water-Resources Investigations*. These publications tend to be of broad scientific or geographic interest and usually focus on a specific hydrologic or geologic topic. Almost all of the publications in these series are available to the public for purchase; *Circulars* and *Fact Sheets* are free of charge. *Professional Papers* document reports pertaining to resource studies for professional scientists and engineers. USGS *Bulletins* contain scientific research (data collected and/or interpreted) pertaining to a geographic area, or are published anthologies of a specific subject matter. Some *Bulletins* are only available online, whereas others are available for purchase (U.S. Geological Survey, 2000c). *Water-Supply Papers* consist of reports on interpretations of various aspects of hydrology and hydrogeology. (The *Water-Supply Paper* series was discontinued in 1996, although papers already in the review process will be published in the series.) USGS *Circulars* contain both technical and nontechnical information on a variety of subjects, such as details of USGS programs and policies. Field manuals for research projects and programs of the USGS are an example of the type of material published in the *Techniques of Water-Resources Investigations* series. Overall, this series provides descriptions of technical procedures and analysis methods used by the USGS.

The USGS also produces other publications to inform the public and geoscientists about ongoing and completed projects. *Fact Sheets* are one type of information product available free to the public, and the publications are intended to provide timely summaries of and results from current and ongoing geoscience investigations. In the informal reports series, *Water-Resources Investigations Reports* (*WRIRs*) provide results and interpretations from hydrologic studies of local or regional scope. *Open-File Reports* (*OFRs*) present geoscience data, preliminary results of investigations, documentation for programs and projects done by the USGS, and miscellaneous information that may not be suitable in other report series. Although most *WRIRs* and *OFRs* are available through the U.S. Geological Survey Information Services Branch,

TABLE 2a. National Information Sources for U.S. Series

USGS Catalogs and Sources for Online	
INFORMATION SOURCE	
Catalog of U.S. Geological Survey Publications (listed by Month and Publication Type)	*http://pubs.us*
U.S. Geological Survey Formal Reports	
Professional Papers (1902-present)	*http://greenw*
Bulletins (1883-present)	*http://greenw*
Water-Supply Papers (1896-1997)	*http://greenw*
Circulars (1933-present)	*http://greenw*
Thematic Maps (1900-present)	*http://greenw*
USGS Registered On-Line Water-Resouces Reports	*http://water.u*
Ground-Water Atlas of the United States	*http://capp.wa*

TABLE 2b. National Information Sources for U.S. Series

USGS Catalogs and Sources for Online	
INFORMATION SOURCE	
Online Professional Papers	*http://geology*
Online Bulletins	*http://geology*
Online Water-Supply Papers	*http://geology*
Online Circulars	*http://geology*
Online Open-File Reports	*http://geology*
Online Fact Sheets	*http://geology*
Online Fact Sheets	*http://geology*
Online Poster Sessions (from professional meetings)	*http://geology*
Online General Interest Publications	*http://geology*

TABLE 2c. National Information Sources for U.S. Geological Survey Publication Series

USGS Publication Catalogs, Indexes, and Bibliographic Databases	
INFORMATION SOURCE	**WEB ADDRESS**
National Geologic Map Catalog	U.S. Geological Survey: *http://ngmdb.usgs.gov/ngmdb/ngm_catalog.ora.html*
Publications of the U.S. Geological Survey	Subset of the GeoRef Database–American Geological Institute, Inc via Community of Science, Inc. *http://usgs-georef.cos.com/*
Selected Water Resources Abstracts	U.S. Geological Survey: *http://water.usgs.gov/swra*
U.S. Geological Survey Library System	*http://www.usgs.gov/library*
Water Resources Scientific Information Center Database	U.S. Geological Survey and the Southern Illinois University *http://www.uwin.siu.edu/databases/wrsic/*

these reports usually also are available through the USGS office that prepared the publication. To search for a USGS hydrologic publication by geographic area, one should use the publications list, bibliography, or searchable bibliography available from each of the USGS WRD offices in the geographic area of interest. (See examples for Atlantic coast offices in Tables 1a, 1b, and 1c.)

Most USGS hydrologic reports and other hydrologic information are produced for a local geographic area in conjunction with other cooperating federal, state, and local agencies. These publications (typically *WRIRs* and *OFRs*) provide local agencies with readily available publications on availability, use, and quality of local or regional water resources. Although citations and abstracts are indexed for all publications, only a fraction of the publication information is accessible at a regional, national, or commercial level (for example, bibliographic databases available through government agencies or commercial vendors). Time lapses between the distribution and indexing of the local publications in national and commercial databases can help explain the limited inclusion of the latest hydrologic citations from lower levels. A general trend of publication and citation availability of USGS publications in hydrology is illustrated in Figure 2.

In summary, a researcher or interested layperson seeking publications on a local hydrologic topic will be best served with the use of alternative sources of information available through the state and program offices of the USGS, in addition to the sources available through state agencies and other federal agencies. Many publications are published, available, and indexed primarily at the local level, through offices such as the USGS WRD state offices (Figure 2). Literature

retrievals conducted through the use of only con
such as the *GeoRef* or the GPO interface availabl
brary Center, Inc. (OCLC) *FirstSearch*, will limi
graphic information obtained. Use of the bibliogra
provided at a local level through state represent
result in additional relevant information retrieva

Alternative Sources of Bibliographic Informa Available from the U.S. Geological Survey

Although the formal and informal reports are
series, the publication and dissemination policies
pline, as well as among the division centers and s
cation practices remain within the general guideli
the USGS, part of the process associated with t
publications is established at a local level (for ex
sometimes by projects). This establishment crea
bibliographic information by specific geographic

For the bibliography of saltwater intrusion a
the online publication lists and bibliographies o
the Atlantic coast were reviewed for relevant in
1c). The results of the online searches were u
publications produced by these WRD state offic
tion lists and bibliography pages, indicating that
cess to the bibliographic information for inte
geographic area or topical discipline. In many c
tion lists and/or bibliographies, the offices pro
federal agencies related to water resources, en
environmental protection. Searchable bibliogra
USGS WRD offices in Maryland-Delaware-Di
and North Carolina. Subject inquiries can be
graphic sources through the use of keywords,
descriptors list of the USGS's Water Resourc
Survey, 1980). The Florida USGS offices ma
(Tables 1a, 1b, and 1c) that specifically addres
water intrusion (saltwater encroachment). This
example of how USGS WRD state-level offic
the public to literature that pertains to a specific

Bibliographies and publication lists provide
and other offices provide invaluable citation inf
and topical categories; however, if one is not
hydrologic topic, the sources indexed at the na

an appropriate location to begin a bibliographic search. Some additional sources of bibliographic information (Tables 2a, 2b, and 2c) include the USGS Library, *Publications of the USGS*, *USGS Selected Water Research Abstracts*, the *Water Resources Scientific Information Center Database*, and the *National Geologic Map Catalog*.

STATE AND OTHER FEDERAL PUBLICATION SOURCES

Sources from State Geological Surveys and Water-Resources Agencies

State geological surveys and water-resources agencies provide useful sources of literature information on local data collection and hydrological investigations (Tables 3a and 3b), most of which are disseminated and indexed at local and regional levels. Often, state agencies collaborate with federal agencies on hydrologic and geologic investigations and publications within a geographic area. As is the case for USGS publications, bibliographic sources for state documents in hydrology are sometimes difficult to obtain through reference databases and library catalogs. Some online bibliographic sources produced by these agencies focus on hydrologic publications, such as the online bibliographies of state geological surveys that were examined for the saltwater-intrusion bibliography and listed in Tables 3a and 3b.

Hydrological and hydrogeological investigations are an important part of the work done by many state geological surveys, and the results are subsequently published by the surveys. The publication lists provided by each state geological survey offices along the Atlantic coast are summarized in Tables 3a and 3b. In some states, the access to geoscience information is through another state office. For example, in New York State, the state geological sources are available through the New York State Museum. Similarly, the Rhode Island state geologist at the University of Rhode Island provides information at the state level. State bibliographic sources are available from agencies in the form of citation lists or publication sales lists. Because the publications originate from the state agencies, the state geological surveys provide more access points to the bibliographic information than do the state depositories. Depositories do not necessarily catalog all state and federal publications, especially gray literature in hydrology. State academic and federal library catalogs, however, offer valuable information and serve as a collection point for the compilation of citation information into commercial and non-commercial bibliographic databases.

Academic and Government Library Catalogs

Library catalogs available from academic and government institutions provide excellent sources of bibliographic information (Table 4). In addition to

providing the library community with access
tions, library catalogs also function as electronic
tion. The monograph and serial citations prov
correlated with the database providers, such as
like *WorldCat* users can obtain the library hold
For publications in hydrology, the information
often derived from the USGS Library's catalog
ernment catalogs. For example, nearly all USG
Open-File Report series have been cataloged
available through OCLC or on the Internet (Ta

Searching academic and government catalogs
for USGS reports published in cooperation wit
brary catalogs for citation verification of USGS a
compile the saltwater-intrusion bibliography, va
ings were identified. In some cases the main en

TABLE 3a. Information Sources for State Geo
sources Offices for States on the Atlantic Coas

State Geological Survey and Web Source	
Connecticut Geological Survey *http://dep.state.ct.us/cgnhs/index.htm*	*Use the Connec* *www.cslib.org*
Delaware Geological Survey *http://www.udel.edu/dgs/dgs.html*	*Online Publicati*
	Publication Cata (bibliography of
Florida Geological Survey *http://www.dep.state.fl.us/geo*	*List of Publicatio* *http://www.dep.s*
	Online Publicatio *http://www.dep.s*
Georgia Geological Survey Branch	*http://ganet.org/* (contact informati
Maine Geological Survey *http://www.state.me.us/doc/nrimc/mgs/mgs.htm*	*Main Publicatior* *http://www.state.*
	Water Resource *http://www.state.*
Maryland Geological Survey *http://mgs.dnr.md.gov*	*Online Pamphle* *http://mgs.dnr.m*
Massachusetts Office of Environmental Affairs	*http://www.magr*
New Hampshire Geological Survey	*http://www.des.s*

TABLE 3b. Information Sources for State Geological Survey and Water Resources Offices for States on the Atlantic Coast

State Geological Survey and Web Source	Source Descriptions
New Jersey Geological Survey *http://www.state.nj.us/dep/njgs/index.html*	*New Jersey Geological Survey Reports and Documents* *http://www.state.nj.us/dep/njgs/pricelst/njgsrprt.htm*
New York Geological Survey *http://www.nysm.nysed.gov/geology.html*	*Publications and exhibits* (New York State Museum)
North Carolina Geological Survey *http://www.geology.enr.state.nc.us*	*Bibliographic information and annotations* *http://www.geology.enr.state.nc.us/bibliogr.htm*
Pennsylvania Geological Survey *http://www.dcnr.state.pa.us/topogeo/indexbig.htm*	*Pennsylvania Geological Publications* (PDF file) *http://www.dcnr.state.pa.us/topogeo/pub/lop98.pdf*
	Popular Publications (available free upon request) *http://www.dcnr.state.pa.us/topogeo/pub/popular.htm* *Publications Page* (includes online publications) *http://www.dcnr.state.pa.us/topogeo/pub/pub.htm*
Rhode Island Geological Survey *http://www.uri.edu/cels/gel/rigs.html*	*Paper Maps for the State of Rhode Island* (includes bibliography for the ground-water map series) *http://www.uri.edu/cels/gel/papmaps.htm*
South Carolina Geological Survey	*http://www.dnr.state.sc.us/geology/geohome.html*
Virginia Department of Mines, Minerals, and Energy	*Division of Mineral Resources* *http://www.mme.state.va.us/Dmr/PUB/publist.html*

TABLE 4. Selected Government and Academic Library Catalogs

Library	Online Access to Catalogs
Library of Congress	*http://www.loc.gov*
National Oceanic and Atmospheric Administration Library	*http://www.lib.noaa.gov*
University of Florida Library	*http://web.uflib.ufl.edu*
University of Rhode Island Library	*http://library.uri.edu/screens/opacmenu.html*
U.S. Department of the Interior Library	*http://library.doi.gov*
U.S. Geological Survey Library	*http://www.usgs.gov/library*
U.S. Department of the Interior Natural Resources Library	*http://www.doi.gov/nrl*

was the issuing agency. Similarly, the report serie *per* or *USGS Open-File Report*) varies between b or only searchable as an added entry. The discrep trieval methods of a search are a result of the dive as well as varying opinions of the individual cata

Different terminology used by providers of also affect how researchers retrieve relevant bibl ample, the preferred word choice for the subject ' tween what is suggested in the USGS's *Water R* and in the *Library of Congress Subject Headings* in other thesauri available from commercial and databases (Table 5d).

COMMERCIAL AND NON-CO BIBLIOGRAPHIC DATA

Comparative evaluation of bibliographic dat an understanding of the coverage, frequency of indexing practices of databases, which can aid i plicable sources to use for a particular literatur hydrologic literature that pertains to saltwater coast, databases accessed through commercial organizations were used to help obtain an all-inc bases evaluated included: *GeoRef* (Cambridge munity of Science, Inc.); *Water Resources Ab* Abstracts, Inc.); *U.S. Government Printing Offi*

TABLE 5a. Selected Keywords and Subject Hea

Selected Keywords (Descriptors), Used for the Saltw from the U.S. Geological Survey *Water Re*		
Descriptor(s)	**Term Used For:**	**Relat**
Saline water	Salt water	Brines Seawat
Saline water barriers	--	Groundw
Saltwater intrusion	--	Encroac
Saline-freshwater interfaces	--	(narrowe

TABLE 5b. Selected Keywords and Subject Headings for Saltwater Intrusion

Selected Subject Headings from the Library of Congress
(Library of Congress, 2000a, 2000b)

Descriptor(s)	**Term used in place of:**	**Related Term(s)**	**Broader Term(s)** **[Narrower term(s)]**
Seawater	Sea-water	--	Saline waters
Saline water barriers	Barriers, Saline water Freshwater barriers Groundwater barriers Saltwater barriers Seawater barriers	--	[Saltwater encroachment]
Saline waters	Waters, Saline	--	[Brackish waters] [Saltwater encroachment]
Saltwater encroachment	Encroachment, Saltwater Intrusion, Saltwater Saline water intrusion Saltwater intrusion Seawater encroachment Seawater intrusion	--	Saline water barriers Saline waters

TABLE 5c. Selected Keywords and Subject Headings for Saltwater Intrusion

Selected Subject Headings from the Library of Congress
(Library of Congress, 2000a, 2000b)

Descriptor(s)	**Term used in place of:**	**Related Term(s)**	**Broader Term(s)** **[Narrower term(s)]**
Geology	Geoscience	--	[Hydrogeology]
Groundwater	Ground water	Hydrogeology	Water
Hydrogeology	Geohydrology	Groundwater	[Aquifers]

Saltwater intrusion USE Saltwater encroachment

Seawater barriers USE Saline water barriers

Seawater encroachment USE Saltwater encroachment

Seawater intrusion USE Saltwater encroachment

Saltwater barriers USE Saline water encroachment

Groundwater barriers USE Saline water barriers

TABLE 5d. Selected Keywords and Subject He

Selected Keyword Search from Cambridge Sci *GeoRef* Database and *Water Resources A*	
Descriptor(s)	**Term used in place of:**
***GeoRef* Database**	
Salt-water intrusion	Encroachment (Ground water) Intrusion (Ground water) Salt water intrusion Sea-water encroachment Sea-water intrusion
Ground water	Groundwater Underground water
Hydrogeology	Geohydrology
***Water Resources Abstracts* Database**	
Saline water intrusion	Encroachment
Groundwater	Phreatic water Subterranean water
Hydrogeology	Geohydrology

Publications of the USGS (Community of Scien *sources Abstracts* (USGS); *National Geologic* the *Water Resources Scientific Information C* Southern Illinois University). The databases var drology and sources of materials indexed; curren the searchable fields; and use of controlled ter among the databases were the search-feature ca ture, and the presence or absence of USGS publ

For each searchable bibliography, bibliographic combination of search strategies was used to retriev These strategies, combined with proper terminolog ter-intrusion bibliography and should result in a cc trieval for those searching the same sources for standard search strategies included the use of con language. A controlled-vocabulary search is the us ject fields for the database or catalog–similar to th

TABLE 6. Selected Bibliographic Databases from Commercial Vendors and Non-Commercial Organizations

Database–Provided by:	Available through:
Commercial Vendors	
GeoRef–American Geological Institute, Inc.	Community of Science, Inc. (COS) Cambridge Scientific Abstracts (CSA)
Water Resources Abstracts	Cambridge Scientific Abstracts (CSA)
Government Printing Office (GPO) Monthly Catalog	Online Computer Library Center, Inc. (OCLC)
Non-Commercial Organizations	
Publications of the U.S. Geological Survey *http://usgs-georef.cos.com*	Subset of the *GeoRef* Database–American Geological Institute, Inc. via Community of Science, Inc.
Selected Water Resources Abstracts	U.S. Geological Survey: *http://water.usgs.gov/swra*
National Geologic Map Catalog	U.S. Geological Survey: *http://ngmdb.usgs.gov/ngmdb/ngm_catalog.ora.html*
Water Resources Scientific Information Center Database	U.S. Geological Survey and the Southern Illinois University *http://www.uwin.siu.edu/databases/wrsic*

the USGS *Water Resources Thesaurus* (Table 5a) or the *Library of Congress Subject Headings* listed in Tables 5b and 5c (Library of Congress, 2000a, 2000b). A natural-language search is the use of everyday terms entered or combined in a variety of ways. Natural-language terms can be truncated (for example, salt* would retrieve all words with salt and the suffixes), used with Boolean logic (for example, [salt not saline] and water and [encroachment or intrusion]), or used with proximity features (for example, salt(3w)intrusion would require intrusion to be within three words after salt), as recommended by Harter (1986). Controlled vocabulary and natural language both have advantages and disadvantages. Because the purpose in searching the electronic bibliographic sources was to compile and verify citations for the saltwater-intrusion bibliography, all of the various search techniques were used.

Commercial Bibliographic Databases

GeoRef is a bibliographic database for the earth sciences that is produced by the American Geological Institute, Inc. (AGI), and is available through the vendors Community of Science, Inc. (COS) and Cambridge Scientific Abstracts (CSA). Since 1966, AGI has been collecting and indexing bibliographic information published since 1785 for the geosciences (American Geological Institute, 2000). Every year, approximately 70,000 *GeoRef* database, adding to the more than 2 mill ords (American Geological Institute, 2000). To trievals can differ between two providers of the s through COS and CSA was evaluated for c searched, as recommended by Harter (1986). [*G* vendors such as Dialog, NERAC, Questel-Orbit, C and STN International.]

GeoRef, as provided through COS, contain geoscience Masters' theses and doctoral dissert ticles and conference proceedings (Communit this database are able to choose from three searc ject search, or advanced search) or by the source which includes the coverage of USGS and state search interface allows the user to search by su first and/or last name, report number, volume year, abstract, keywords, affiliation name, conf tion, language, document type, or map type. Th search the descriptors in the database and then terface. The advanced search enables the user to mation by a specified field. The source search al journals and report series indexed in the databa search interface. In the COS *GeoRef* database, "U.S. Geological Survey" affiliation as of May search options and features, an author index and with the option to continue to the main search i list of authors' citations, including variations o ple, Mountain, Kate; Mountain, K.; Mountain, The keyword index is a list of preferred descript database terms for saltwater intrusion, which di lined in the *Water Resources Thesaurus* (198 (392 records) and "encroachment (groundwate were performed by entries into the descriptors intrusion" retrieved 799 records, and the "enc ords. The discrepancies between these record re in all cases the searches performed were word s water and intrusion." For each search, the searc present anywhere in the bibliographic record Nonetheless, this example demonstrates how different citation retrievals.

Search strategies and results will vary within each platform of *GeoRef* because of the database retrieval structure and search features. In the *GeoRef* provided by CSA, users are able to truncate natural language terms and use Boolean logic within keyword, title, author, or journal name in a standard search mode, as well as to specify the publication year (1960 to 2000). The CSA *GeoRef* database also provides the user with the option to display the complete bibliographic record, the citation, and/or abstract search results by retrieval relevancy or date of distribution. In the advanced search option, users can search for bibliographic records by terms in the fields: keywords, author, title, abstract, descriptor, source, agency, author affiliation, corporate author, conference, classification, editor, entry month, environmental regime, identifiers, input center, or international standard number. According to the information page of the *GeoRef* files, the database is updated weekly (Cambridge Scientific Abstracts, 2000). Unfortunately, at the time of the evaluation, no thesaurus was available for this version of *GeoRef* provided by CSA. Therefore, terms were applied from the *Water Resources Thesaurus* and selected subject headings from the Library of Congress, outlined in Table 6.

The sample searches were not limited to the problem of saltwater intrusion within the United States, or specifically to the Atlantic coast, because the use of the geographically unrestricted search terms resulted in a manageable number of retrievals. When the keyword search option was used, the terms "saltwater intrusion" (165 records), "saltwater encroachment" (20 records), and "saline encroachment" (3 records) resulted in additional and precise retrievals. In comparison, the same terms were used for searches in the descriptor fields and resulted in a lower recall of relevant records: "saltwater intrusion" (2 records), "saltwater encroachment" (1 record), and "saline encroachment" (1 record). To analyze the coverage of USGS publications, the source and author affiliation were searched. At the time of the assessment, the database contained 76,789 records of USGS publications (for example, 30,022 Open-File Reports; and 14,734 Professional Papers), and 30,866 records included USGS authors.

Non-Commercial Bibliographic Databases

Non-commercial bibliographic databases can be easily accessed electronically through government agencies (such as the USGS), universities, and other organizations that provide large bibliographic databases at no cost. The non-commercial databases used for the saltwater-intrusion bibliography, which are relevant to hydrology in general, include the *Publications of the U.S. Geological Survey*, *Selected Water Resources Abstracts*, *National Map Catalog*, and the *Water Resources Scientific Information Center Database* (Table 6). The *Publications of the U.S. Geological Survey* database is provided through COS, where the search options are the same as those the previous section. This is the most recent se public for retrieval for bibliographic informati database includes articles by USGS employees the USGS. Another database currently availabl the *National Geologic Map Catalog*. Although yet complete, it provides users with citation in tial publications (for example, the thematic through the USGS National Mapping Disciplin

Selected Water Resources Abstracts (*SWRA* USGS publications relating to hydrology that is ter Resources Discipline main Web page. This *Water Resources Abstracts*, the commercial da Though *SWRA* is available at no cost and is ea pages, it should be noted that it is "selected," ra neither inclusive nor current, but it is nonethel historical hydrologic bibliographic informatio saltwater-intrusion bibliography. *SWRA* prov 1977-93 for literature in which the corporate a publications and literature published by USGS a logical Survey, 2000d). [An internal USGS bib *Report Tracking System (RTS)*–created in 1997 collection of citation information.] One of the database offers is the option to retrieve docume (HUCs). This feature allows the retrieval of bibl basin as a result of entering the 8-digit HUC cat ment in the hierarchy of hydrologic units" (U.S. example, if one wanted to retrieve all of the bibl ing to the Blackstone River Basin in Massach *SWRA* database may be searched using "0109 USGS *Water Resources Scientific Information* vided by the University Water Information Netw the complete collection of abstracts from 1977 2000d). The UWIN database can be accessed usgs.gov/swra/), or directly (http://www.uwin.s

CONCLUSION

Access to bibliographic sources depends on kr is located as much as on one's physical access t

tional online bibliographic databases and indexes provide access to materials in hydrology; however, these tools do not provide comprehensive coverage of relevant hydrologic publications. This paper describes electronic (online) sources of bibliographic information that may be useful to hydrologists, librarians, and others interested in hydrology. These sources are (1) bibliographic information available through the U.S. Geological Survey, (2) references of hydrologic publications from other agencies of the federal government, (3) state geological surveys and water-resources agencies, and (4) state depositories. Many of the online sources were used while compiling a bibliography on the subject of saltwater intrusion along the Atlantic coast of the United States. The paper also provides an overview of the types and distribution structure of hydrologic publications of the U.S. Geological Survey.

A more universal bibliographic indexing system than currently present is suggested to indexers and library professionals to provide more efficient organization of citation retrievals. With the increasing use of online publication lists and bibliographies, researchers and administrators may also consider bibliographic sources to document online versions of the information. Currently, the most effective way to retrieve publication information, such as for the Atlantic coastal bibliography, is primarily through colleague referrals and through the bibliographies and publication information from USGS state offices and state geological surveys.

REFERENCES

American Geological Institute. *GeoRef Information Services*, 2000, <http://www.georef.org> (30 April 2000).

Barlow, P.M. *Ground-Water Resources For the Future: Atlantic Coastal Zone*. U.S. Geological Survey Fact Sheet 085-00. Reston, VA: U.S. Geological Survey, 2000.

Barlow, P.M., and Wild, E.C. *Bibliography on the Occurrence and Intrusion of Saltwater in Aquifers along the Atlantic Coast of the United States*. U.S. Geological Survey Open-File Report 02-235. Northborough, MA: U.S. Geological Survey, 2002.

Bichteler, Julie. "Geologists and Gray Literature: Access, Use, and Problems." *Science & Technology Libraries* 11 (Summer 1991): 39-50.

Bichteler, Julie, and Ward, Dederick. "Information-Seeking Behavior of Geoscientists." *Special Libraries* 80 (Summer 1989): 169-78.

Cambridge Scientific Abstracts. *Brief File Descriptions: Environmental Sciences*, 2000, <http://www.csa.com/siteV3/ids-envi.html> (3 May 2000).

Community of Science, Inc. *About GeoRef*, 2000, <http://georef.cos.com/About_georef.shtml> (30 April 2000).

Harter, S.P. *Online Information Retrieval: Concepts, Principles, and Techniques*. San Diego, Academic Press, 1986.

Library of Congress, Cataloging Distribution Servi
Headings. Vol. 2, D-H. Washington, D.C., 2000a
________. *Library of Congress Subject Headings*. V
U.S. Geological Survey. *Water Resources Thesauru*
________. *Mission of the USGS*, 2000a, <http://www
(13 February 2001).
________. *Mission of the Water Resources Divi*
gov/wrd_mission.html> (15 July 2000).
________. *New Publications of the U.S. Geological S*
tions Issued January-March 2000. Reston, VA, 2
________. *Instructions for Searching Selected Wa*
<http://water.usgs.gov/swra/help.html> (24 April
________. *Hydrologic Unit Maps*, 2001 <http://w
August 2001).

The Latest and the Best: Information Needs of Pharmacists

Mignon Adams

SUMMARY. Pharmacists were once primarily dispensers of drugs but are now considered the drug information member of a health care team. They thus need the most recent and the most accurate information available on drug therapeutics. Librarians can help pharmacy students acquire the skills they need in order to locate the best evidence available. *[Article copies available for a fee from The Haworth Document Delivery Service: 1-800-HAWORTH. E-mail address: <docdelivery@haworthpress.com> Website: <http://www.HaworthPress.com> © 2001 by The Haworth Press, Inc. All rights reserved.]*

KEYWORDS. Pharmacy, pharmacists, drug information

INTRODUCTION

Most people see a physician or a nurse no more frequently than several times a year, yet they are generally aware of the education needed for these

---

Mignon Adams, BS, MLS, is Director of Library and Information Services, University of the Sciences, Philadelphia (formerly the Philadelphia College of Pharmacy and Science), 600 South 43rd Street, Philadelphia, PA 19104 (E-mail: m.adams@usip.edu).

The author gives appreciation to Dr. George E. Downs, Dean of Pharmacy at the Philadelphia College of Pharmacy, University of the Sciences in Philadelphia, for reviewing and commenting on early drafts.

[Haworth co-indexing entry note]: "The Latest and the Best: Information Needs of Pharmacists." Adams, Mignon. Co-published simultaneously in *Science & Technology Libraries* (The Haworth Information Press, an imprint of The Haworth Press, Inc.) Vol. 21, No. 3/4, 2001, pp. 87-99; and: *Information and the Professional Scientist and Engineer* (ed: Virginia Baldwin, and Julie Hallmark) The Haworth Information Press, an imprint of The Haworth Press, Inc., 2001, pp. 87-99. Single or multiple copies of this article are available for a fee from The Haworth Document Delivery Service [1-800-HAWORTH, 9:00 a.m. - 5:00 p.m. (EST). E-mail address: docdelivery@haworthpress.com].

http://www.haworthpress.com/store/product.asp?sku=J122

10.1300/J122v21n03_06

professions and know what each is qualified to d teract with a pharmacist each time they have a pr little or no idea what a pharmacist does, the edu need for information.

Pharmacy as a profession has much in comm brarians, pharmacists are called by the name of they do; anyone behind a pharmacy counter is and most people think that little specialized educ do the job. And also like librarians, pharmacis transformed over the past twenty years as the in migrated from print to electronic.

The most common entry-level degree for phar the Doctor of Pharmacy, or Pharm.D, a clinical of six years of higher education: two years (mostly in the basic sciences), followed by four tion. Pharmacists who wish to specialize will t residency or fellowship in an area such as cardio tion, or pediatric or geriatric drugs.

There are only 83 pharmacy schools in the cou medical schools. Though 7,000 pharmacy degree there are fewer pharmacists than jobs available. acute in retail stores. The average starting salary was reported to be over $65,000[2] (although it shou increase with experience as quickly as this may h

When Americans think of a pharmacist, they p retail pharmacy. However, while there are an e pharmacists in the country, only 130,802, or 38% stores.[3] The next largest group of pharmacists is and consultant pharmacists, those who work in form a variety of duties in hospitals. Consultant p home residents and others who are in long-term cists are hired to work in HMOs, pharmaceutica agencies such as the U.S. Public Health Service a tion. All these various roles require knowledge they are used in therapy.

Although the common image of the pharma penses drugs, pharmacy is now becoming a pr products, but with information. A pharmacist ne most accurate information about drug therapy ar Pharmacists use this information in helping to n tient care, creating formularies for the use of lar

sionals, translating and integrating drug information for those in other health professions, and educating patients so that they understand their medications and make the best use of them.

In addition, graduates with a pharmacy degree may choose advanced degrees in such areas as pharmacology (the study of how drugs work in the body), pharmaceutics (how drugs are delivered to the body), medicinal chemistry (the design of new drugs), or pharmacoeconomics (cost-benefit and cost-analysis of drug therapy). However, this article will consider the information-gathering behaviors of pharmacists who work as pharmacists. Eleven pharmacists were interviewed for this article, including academic, hospital, retail, and industrial pharmacists. Their comments are integrated into the following discussion.

INFORMATION NEEDS OF PHARMACISTS

Retail/Community Pharmacists

Many may have fond memories of "Doc" at the corner drugstore, who owned and ran his (seldom "her") business. He may have lived nearby, and been available late at night and on Sunday for sudden fevers or other health concerns. His customers came to him not only to have their prescriptions filled, but also for advice on the best over-the-counter medication for a sore throat or even to have a mote removed from an eye. Doc's drugstore may have long ago been purchased by a chain, but there are still 20,000 independent community pharmacies, filling about 20% of all prescriptions.[4]

About 40% of prescriptions are now filled by large chain pharmacies, whose size and hours may rival the local supermarket. An additional 20% are filled by pharmacies located in supermarkets or large merchandisers like Wal-Mart or K-Mart. The final 20% are filled by mail-order pharmacies.[5] But at all these community pharmacy sites, the role of the pharmacist is the same. He or she is responsible for overseeing the dispensation of drugs (often actually carried out by a technician), being available to counsel patients about the drugs they take and the importance of taking them, and watching for possible drug interactions or side effects.

In chain drug stores, much of this information may be contained in the computer database compiled and maintained by pharmacists located elsewhere. Within the database are instructions for the patient, written in lay language, about dosage requirements and possible side effects. Each customer record contains all the drugs the customer has received from that chain, so that possible drug interactions can be spotted.

In addition, the pharmacist needs to be well i
ter drugs, nutritional supplements, and more re
These substances may interact with each other o
tomers need help in selecting the best one for t

Many community pharmacists have receive
they can administer vaccinations or monitor dr
agement of high cholesterol, hypertension, coa
Each of these therapies requires the monitoring
education of patients as to the importance of co

Hospital and Consultant Pharmacists

The 38,000 hospital pharmacists and consul
nursing homes and other long-term care institu
responsible for the dispensation of drugs to pa
macists are. But they also have other important
in the development of *drug formularies*. A form
been reviewed for their safety, effectiveness, a
mulary thus requires knowledge both of curre
relative costs of similar therapies. The formula
professionals to guide their selection for indivi
may influence the inclusion of a drug more tha

Since a bedridden patient cannot come to a
institutional pharmacist must oversee a delive
drugs arrive at the patient's bedside quickly and
consultant pharmacists also are responsible for
venous drugs. Some pharmacists head up com
IV drugs to homebound patients.

Another important role for pharmacists is tha
fessionals on new drugs and new applications a
to them as needed. Specific drug information qu
U.S. equivalent for a drug prescribed in anoth
brought in by a patient; determining which dr
unwanted side effect; or providing very quickl

In a large teaching hospital, the roles des
among several different pharmacist groups. T
macy, where a pharmacist oversees the filling o
or a robot. There may be a separate drug inform
bers prepare the formulary and respond to queri
be pharmacists who accompany physicians on
mediate resource. These pharmacists may respo

ations, so that they can provide the right drug at the correct dosage and allow the physicians and nurses to concentrate on saving the patient's life.

In a nursing home or other long-term care facility, the pharmacist is responsible for reviewing all the drugs that each resident takes, taking note of appropriate administration, possible duplications, and interactions with other drugs or food. Since most residents in these facilities are elderly, the pharmacist must be knowledgeable about the use of drugs with geriatric patients and their particular effects on the aged. The consultant pharmacist may also supply medical devices and must be familiar with their use and appropriateness.

Pharmacists in the Pharmaceutical Industry

Pharmacists fill many jobs in pharmaceutical companies. One of these is to staff the companies' drug information centers, answering questions from physicians and other pharmacists. They must be very well informed about each drug their company produces and all pertinent clinical studies. Pharmacists also help conduct clinical trials. In this role they must not only be familiar with relevant clinical studies, but also with the regulatory requirements of the Food and Drug Administration (FDA).

Pharmacists, along with nurses and others trained in the sciences, are often pharmaceutical sales representatives. However, they may also be "medical science liaisons," whom the FDA allows to speak not just on the approved uses of their company's drug, but also for "off-label" uses that have been reported in the medical literature. In addition to knowledge about business and marketing, these pharmacists, like others, must be familiar with the clinical medical literature.

HOW PHARMACISTS GATHER THE INFORMATION THEY NEED

In the many different roles that pharmacists play, the need for current medical literature is paramount. How they meet that need has changed profoundly in the past decade as they have increasingly taken advantage of the electronic tools available to them.

Clinical Studies

The "gold standard" for evidence is the controlled randomized trial, reported in any one of a number of medical journals. Before the Internet, a common method used by many pharmacists to keep up was to skim the tables of contents of a number of different journals. One pharmacist reported that she previously scheduled a two-day session every six months at a medical library,

perusing the tables of content from many journ
cles she thought would be helpful. Now, wit
bles-of-content services, she receives e-mails v
published.

Other electronic alerting services are now
American Society of Health-System Pharmacist
pharmacists of articles of interest to them. We
specialty organizations also often have alerting

When a pharmacist wants to retrieve many a
or she turns to *Medline*, the world's largest da
ences. Provided free of charge by the National L
product *PubMed* to anyone in the world, this has
practitioners, those in small organizations and
countries. *PubMed* also provides an easy metho
cal studies. The filter, "Clinical Queries," autc
headings ("randomized controlled trial," "plac
that the results are limited to studies that are like

Embase, the Elsevier biomedical science da
Japanese research literature more thoroughly th
high cost, particularly compared to no cost for P
employed at large pharmaceutical firms are lik

Review Articles

Review articles summarize the research in a
review article is most likely to summarize studi
for a specific disease or condition. However, ove
view articles have indicated that they tend to be
ner using procedures that cannot be replicated.

"Systematic reviews" are review articles in w
lecting those articles he or she thinks are most re
tocol. First, there is an exhaustive search of th
least two different databases each searched by
following predetermined criteria, those articles
sound are selected. Finally, conclusions are dra
meta-analysis is done, then the researchers not c
sound studies, but also analyze the results so th
sizes can be combined to produce more significa
ical topic, it is possible to find studies that contra
views and meta-analyses are methods develope
conflicting conclusions.

Medline searches can be so constructed so as to identify both systematic reviews and meta-analyses. In addition, there are specific databases of these articles, including *Best Evidence*, the *Cochrane Database of Systematic Reviews*, *TRIP* (*Turning Research into Practice*), and *SUMSearch*. However, the pharmacists interviewed for this article have only just begun to use these.

Practice Guidelines

Clinical practice guidelines are systematically developed statements that outline an accepted practice or procedure or treatment for a specific disease or condition. Guidelines should be based on evidence, rather than on general practice, and are usually developed by a medical society or organization. The National Guideline Clearinghouse has a large collection, searchable by either disease/condition or therapy/intervention. Pharmacists use guidelines in developing formularies, since the guideline for, say, systolic dysfunction heart failure specifies the drugs that should be used for that condition.

Drug Monographs

Depending on the source, a drug monograph may be several paragraphs to several pages long. It is a summary of information about a drug, covering its mechanism of action, its therapeutic uses, its side effects, and other pertinent information. Most people are aware of the *Physician's Desk Reference*, which contains drug monographs listed by pharmaceutical company. Since the information in these monographs is no more (and no less) than what the Food and Drug Administration specifies, pharmacists usually prefer to refer to sources with more detail.

There are several collections of drug monographs that pharmacists prefer. One is the *AHFS Drug Information*, prepared specifically for hospital and consultant pharmacists. Each entry is several pages long, and describes the drug's mechanism in detail, including dosage, storage, and preparation information, and much information about side effects. *Drug Facts and Comparisons* groups like drugs together and compares them according to their method of action, formulation (tablets, capsules, liquids, etc.), dosage, and cost. *Drugdex*, a component of *Micromedex*, is highly praised by pharmacists because it links each statement, including therapeutic use, possible side effects or interactions, to the clinical literature.

Textbooks

Pharmacists might refer to textbooks for background information, particularly in an area in which they are not expert. Clinical pharmacists are more likely to refer to a medical textbook than to a pha ical specialty area there are one or two most hig known by the name of the physician who first w may be in many editions past the death of the ori *ciples of Internal Medicine*, for example, is cons tative source on internal medicine. (Its 15th ed named "Harrison.") Textbooks in other specialty "Brandon Hill List."

In the past five years, collections of medic available electronically, so that a drug name, fo across a number of different textbooks simult available only at academic institutions or large

Websites

Several different kinds of Websites are importa of the Food and Drug Administration is perha makes available news on approved drugs and re ters that affect the daily worklife of every pharma turers each maintain Websites that provide a q access the Product Package Insert (PPI), the dru the information the FDA requires to be distribute American Pharmaceutical Association, the Am Pharmacists, the American Society of Health-Sys Community Pharmacists Association, the Natio Stores, and others) have Websites with helpful i

Pharmacists with specialties often access the ily aimed at health practitioners, such as the An or the American Society of Clinical Oncology practice guidelines, and members can usually ac zation's publications.

THE PHARMACIST'S EDUCAT

Pharmacists who graduated ten or more ye methods of keeping up with their informational regularly perused a small number of important yond those journals, they searched the printed in requests for librarians to perform searches for the was a slow process. If they had access to a medic or sent someone to photocopy numerous articles

alerting devices. The Internet has completely changed this process, placing needed access to the clinical literature in the hands of most pharmacists.

With the advent of the Internet, most pharmacists taught themselves to search *Medline* and take advantage of what was suddenly available to them. They also attended workshops held at professional meetings or at a medical library. But since they learned painfully themselves, many pharmacists think that searching techniques can only be obtained by learning on one's own.

Most pharmacy schools require a course in drug information or biomedical literature. In these courses, students are taught to read a professional article and to recognize the importance of the primary literature. They may or may not be taught searching techniques, and if so, it may or may not be taught by librarians.

THE ROLE OF LIBRARIANS

Unlike pharmacists, librarians have had extensive training in searching. In order to locate the current, accurate clinical literature that they need, pharmacists can learn critical skills from librarians. Only pharmacists who work in large pharmaceutical firms, teaching hospitals, or academic institutions will have access to a medical library or librarians, so the skills listed below are those that should be taught as part of the pharmacy curriculum.

Formulating a Proper Search Question

In order to find the kind of evidence that they need, pharmacists must know how to construct a question that is most likely to find methodologically sound studies on the topic they're seeking. Most pharmacists will do their searching themselves. But even when they don't, as for example those who work at large pharmaceutical companies, they need to know how to ask a researchable question.

Using the Powerful Search Features of **Medline**

Medline was one of the first online databases developed, and from its beginning it had a number of powerful features built into it: keyword, Boolean, and subject searching, truncation, "exploding" (the use of hierarchical subject headings), a variety of limits. Most self-taught searchers are unaware that these features exist. Their ignorance of them also means that they often do not understand how they get the results they do.

Using Other Databases

Pharmacists seldom use databases other than *Medline*, even though within the last five years several databases have been developed that make it easier to

find controlled randomized studies and systemati
demonstrate why these databases are useful and t
tabases are constructed so that they understand h

Finding Affordable Access to the Medical Liter

Pharmacists who are employed in academic in pitals, and pharmaceutical companies typically h and journals. Pharmacists in smaller hospitals or consultant positions need to be aware of the res low or no cost and be taught how to use them.

CONCLUSION

"Doc" at the corner drugstore has given way to medical establishment is to serve as the drug infor no longer compound medicines or roll pills. Instea helping to make decisions about the drugs used fo professionals up to date on new drugs and new use tients about the proper use of their medications. great. Their quick adoption of the Internet and its easier, but they need the proficient skills that libr

NOTES

1. American Association of Colleges of Pharmacy
2. *Drug Topics,* http://www.drugtopics.com (9/21
3. National Association of Chain Drug Stores, *The* *file* (Alexandria, VA: The Association, 2002), 43.
4. National Association of Chain Drug Stores, htt
5. National Association of Chain Drug Stores, htt
6. American Association of Colleges of Pharmacy
7. A. Oxman and G. H. Guyatt, "The Science of *the New York Academy of Science 703* (1993): 125-1

APPENDIX

IMPORTANT RESOURCES FOR

Websites

American Pharmaceutical Association (http://w
American Society of Consultant Pharmacists (ht
American Society of Health-System Pharmacists

Centers for Disease Control (http://www.cdc.gov/)

Statistics and news. *Mortality and Morbidity Weekly Report* can be accessed here.

Food and Drug Administration (http://www.fda.gov)

For recent drug approvals, recalls, regulatory changes.

Free Medical Journals (http://www.freemedicaljournals.com)

A labor of love put together by an MD who monitors high quality medical journals to determine which provide some or all of their content free online. Especially appreciated by unaffiliated pharmacists or those at small organizations.

National Association of Chain Drug Stores (http://www.nacds.org)
National Community Pharmacists Association (http://www.ncpanet.org)
National Guidelines Clearinghouse (http://guidelines.gov/)

Source for finding practice guidelines issued by many different agencies.

Pharmaceutical Manufacturers

Manufacturers' websites typically provide the information required by the Product Package Insert (the same information that appears in the PDR).

Indexes and Abstracts

Medline

The National Library of Medicine's database of clinical medicine. Pharmacists access it through *PubMed* or *Ovid.*

Embase

The European database of clinical medicine. Its coverage of European and Japanese research is far better than *Medline*'s. However, its cost (owned by Elsevier) means that it is much less used than it should be.

International Pharmaceutical Abstracts

Produced by the American Society of Health-System Pharmacists. Covers pharmacy and pharmaceutical trade journals as well as clinical journals.

Newsletters and Alerting Services

The Medical Letter of Drugs and Therapeutics (New Rochelle, NY: The Medical Letter, Inc., Monthly)

"Unbiased critical evaluations of drugs v drugs." Highly respected. Slow to adapt to

FDC Reports: "The Pink Sheet" (Chevy Chase, M

News on drug approvals and clinical trial

Tables-of-Contents Alerting Services

Any service, such as Ingenta's Reveal or pu via e-mail tables-of-contents of selected jou

Electronic Newsletters in Particular Areas, e.g.

Pharmacists report that their various special bership in organizations or by subscription, n events. One such is *Health News Daily* (Cl Inc., daily), a newsletter that summarizes leg

Drug Monographs

AHFS Drug Information (Bethesda, MD: Amer Pharmacists, annual)

A publication of drug monographs prepar macists. Especially strong in dosage and

Drug Facts and Comparisons (St. Louis: Facts with monthly updates)

For each therapeutic category (e.g., analg compares action, dosage, and cost. Availa

Micromedex: Drugdex, Identidex, Poisindex (Den quarterly)

An electronic product with many differen of drug monographs prepared by drug inf country. Each statement about the drug's study. *Identidex* is a directory of physical lar tablet, say, can be identified. *Poisindex* thal doses and antidotes. Other sectio *Nominum, Martindale's, Diseasedex*, and

Databases to Locate Systematic Reviews and

The Cochrane Library Online (http://www.upd

Abstracts of the *Cochrane Database of Systematic Reviews* can be searched without a subscription; subscription information is available on the site.

SUMSearch. (http://SUMSearch.UTHSCSA.edu/cgi-bin/SUMSearch.exe)

Free online database maintained by the University of Texas Health Science Center at San Antonio. One search entry retrieves textbook entries and looks for systematic reviews in *Cochrane* and *PubMed.*

Turning Research into Practice (TRIP) (http://www.tripdatabase.com/index.cfm)

Free online database. Funded by Great Britain's National Electronic Library for Health. One search entry retrieves *PubMed* entries that link to free journal articles, practice guidelines, and *Cochrane* and *POEM* systematic reviews.

Other Resources

Basic Medical Textbooks

Harrison's (New York: McGraw-Hill, available in print or electronically) is the standard internal medicine textbook. In addition, pharmacists with a specialty refer to the standard textbook in that area. (See http://www.brandon-hill.com for suggested titles.)

Electronic Collections of Medical Textbooks

Such collections allow the user to search across a number of textbooks with one search entry. Examples are *MD Consult* (Elsevier Science), *StatRef!* (Jackson Hole, WY: Teton Data Systems), and *UpToDate* (Wellesley, MA: UpToDate).

Wall Street Journal (NY: Dow Jones, Inc., daily)

The pharmaceutical industry is one of the largest in the country. If it impacts the industry, *WSJ* reports on it: failed clinical trials, FDA approvals, legislative initiatives.

The San Diego Zoo Library "Began with a Roar"

Linda L. Coates

SUMMARY. The Library of the Zoological Society of San Diego is a curious mix of unique information products, online systems and services. The Zoo Library must cope with unique and difficult-to-obtain information resources, a diverse group of users, and a skeleton staff. Because of the Society's sprawling organizational footprint, a Web-based intranet was established several years ago. Library staff is now attempting to establish a gateway to all of the information resources critical to supporting the Society's plant and animal collections as well as its conservation programs. This article includes selections from the Web portal and the results of a staff survey that attempted to identify the information resources considered most valuable to veterinarians, curators and researchers working in zoos today. *[Article copies available for a fee from The Haworth Document Delivery Service: 1-800-HAWORTH. E-mail address: <docdelivery@haworthpress.com> Website: <http://www.HaworthPress.com> © 2001 by The Haworth Press, Inc. All rights reserved.]*

KEYWORDS. Zoos, conservation, exotic animals, Web portals

The term zoo library has a certain 'oxymoronic' ring to it. Zoos are semi-wild, loosely organized and low-tech while libraries are quiet, orderly and hi-tech. Actually all of these adjectives can be applied to this library–it is a curious pastiche

---

Linda L. Coates, MLIS, is Librarian, Zoological Society of San Diego, P.O. Box 120551, San Diego, CA 92112-0551 (E-mail: lcoates@sandiegozoo.org).

[Haworth co-indexing entry note]: "The San Diego Zoo Library 'Began with a Roar.'" Coates, Linda L. Co-published simultaneously in *Science & Technology Libraries* (The Haworth Information Press, an imprint of The Haworth Press, Inc.) Vol. 21, No. 3/4, 2001, pp. 101-120; and: *Information and the Professional Scientist and Engineer* (ed: Virginia Baldwin, and Julie Hallmark) The Haworth Information Press, an imprint of The Haworth Press, Inc., 2001, pp. 101-120. Single or multiple copies of this article are available for a fee from The Haworth Document Delivery Service [1-800-HAWORTH, 9:00 a.m. - 5:00 p.m. (EST). E-mail address: docdelivery@haworthpress.com].

http://www.haworthpress.com/store/product.asp?sku=J122

10.1300/J122v21n03_07

of systems and services. Zoo Librarians have to ke preserving historic books, finding old citations an of animals (80% of mammals were described be 21st century. They must keep abreast of the latest r cine and research, create institutional databases, a

A research library was proposed in the Zool *corporation* in 1916. Founder Dr. Harry Wegef *Roar*), two other physicians and a naturalist wer mal life in addition to establishing a zoo. They wasn't until 1989 that a professional librarian sists of a single professional (Associate Direct nology Specialist and a Research Specialist. Zoo past decade but there are still only about fifty zo than a dozen of these have more than 5,000 book

Although zoos have existed for centuries, re with zoos are relatively new. They were the resu est in breeding endangered species and managing 1975 The Center for Reproduction of Endanger lished as part of the Zoological Society of San CRES divisions were expanded to include ge physiology, endocrinology, infectious diseases, chemistry, and applied conservation. Collaborati rators, local physicians, and academicians occurr projects in the field were begun. Dr. Alan Dixs University, U.K., was selected to head CRES in 1 ciety's commitment to scientific research and co new postdoctoral fellows to CRES and plans to m field stations in five key areas: China, the sout America and the Pacific and Caribbean islands.

In 1998 The School of Veterinary Medicine nia-Davis (UC-Davis) and the University of Cal tablished the UC Veterinary Medical Center- members from the two institutions to collabora tivities and to offer specialty veterinary service nerships with zoo veterinarians and researcher series of wildlife internships is planned.

Because of these collaborative relationships veterinarians enjoy adjunct professorships at U State University (SDSU). This privilege entitl *views*, *Zoological Record*, *CAB*, *Current Con* number of full-text journals.

THE ZOO'S COLLECTION

The growth of the Library's collection has paralleled the Society's widening focus. Information on conservation, captive breeding and reintroduction are three important areas of expanded interest that have influenced its growth in the past fifteen years. The Library purchases approximately 200 new books each year. The collection has 11,500 monographs relating to exotic wildlife (husbandry, ecology, conservation, and veterinary medicine), horticulture (with an emphasis on palms, aloes, cycads, succulents, tropical and desert plants), and zoos (history, design, and exhibitry). A unique Rare Book Room includes a set of the *Proceedings of the Zoological Society of London* from 1830-1912, Forshaw's *Kingfishers and Related Birds*, Elliot's *Monograph of the Bucerotidae*, and Nardelli's *The Rhinoceros.*

A California State Library Grant enabled the purchase of a file server and software to provide Z39.50 access to the Library Book Catalog at 199.106.195.111. It can also be found at library.sandiegozoo.org. Monographic holdings are available for interlibrary loan through OCLC, under the acronym ZOS.

In addition to LC subject headings, catalog records are enhanced with specialty indexing terms from the *Thesaurus of Zoo and Conservation Terms* that was developed in 1990 to accommodate the Zoo Library's unique scope. An expanded Cutter system facilitates grouping animals taxonomically. (No co-mingling of species in THIS library.) For a more detailed look at what comprises a core Zoo collection, a bibliography (currently under revision) is available at www.sil.si.edu/works/tpbib.htm.

The San Diego Zoo Library currently receives 635 serial publications. They are international in scope reflecting the animal and plant collections, and many are unique to the state of California. Of the 865 titles in the permanent collection, a sizable number would be termed "ephemera," for example, newsletters that are desk-top publications and journals without ISSN numbers that regularly change names, size, and editors. Everything is irregular AND late, and approximately one third of the library's serials must be ordered directly. Some of the more unique holdings include publications from the Wildlife Conservation Union (formerly the International Union for Nature and Natural Resources or IUCN)–a collection of Reports and Action Plans from the Species Survival Commission (SSC), Captive Breeding Specialist Group (CBSG) Reports, and publications from the American Zoo and Aquarium Association (AZA). Guidebooks and annual reports from other zoos, along with studbooks (National/International pedigrees for various species of captive animals) are some of the most unique zoo publications. Although difficult to obtain, they frequently include important bibliographies and valuable husbandry information. These journal holdings are not in OCLC, bu ital Library Database at http://www.melvyl.uco

WEB PORTAL–GEOGRAPHY

Although zoo library users are an extremel PhD scientists at CRES to high-school student front lines, the foremost challenge is not the div ees or the eclectic information resources; it is the Zoological Society of San Diego is spread over occupies 125 acres near downtown San Diego, a 1,800 acres in Escondido. The two sites are sepa ern California Freeway (a formidable barrier as testify). In addition, the Society occupies sever and has a growing number of foreign field work decentralized system of mini-libraries has evolv

In the Main Library, located at the Zoo, the colle onomy, animal behavior, animal husbandry, natur vation. There are two medical libraries, one at ea located. Their collections cover veterinary medici horticulture collection at each site and a large numb mental collections complete the complicated pictu ideal and has resulted in frustrated library staff a Vets, curators and researchers rarely have time to on their mini-libraries, personal article collections E-mail has become the communication mode of c

The complicated geographic distribution of Web-based library system. As soon as the Society Intranet was instituted to function as a front-end p cal information resources. Hyperlinked cross-ref unique resources created by Library staff to pe menu-driven resource is extremely low-tech and varies widely within the organization and the ax should be 90% information and 10% system is st

The information resources that are posted co user suggestions, listservs: Vetlib-L (Veterina AZA-LSIG (Zoo Librarians); alerting services: the Life Sciences, scout.cs.wisc.edu; WildlifeD wildlifedecisionsupport.com; and Peter Dickens elvinhow.prestel.co.uk.

The Portal was designed to accommodate a wide range of browsers, as Society hardware is half-Mac-half-PC. The menus are shallow so that information can be quickly accessed. This feature is useful for staff, who must provide quick answers to the public via the telephone. The opening menu consists of six simple straightforward choices: Animals, Plants, Conservation, Reference, News, and Zoos. The terms: 'books,' 'journals,' or 'indexes' don't appear until an initial selection is made, but these resources are at the top of whatever category is chosen.

SURVEYING SCIENTIFIC STAFF

In August of 2002 the library surveyed the scientific staff to determine how satisfied they were with the information resources available to them. (See Appendix.) The survey was sent to 9 curator/lead keepers, 14 veterinarians and 24 researchers. Seventy-five percent of the surveys were returned within two weeks indicating user support and enthusiasm for library services. In terms of finding information the following search tools were identified by users as most valuable:

NISC's BiblioLine *biblioline.nisc.com*

Wildlife & Ecology Studies Worldwide (Biblioline) is a bibliographic database covering all aspects of wildlife and wildlife management. Less scientific than *Biosis* and *Zoological Record* it includes a healthy quantity of gray literature–much of it relevant to zoos. And unlike most online bibliographic databases, NISC has many valuable older (pre-1935) citations. Best of all, researchers have found that the library has approximately two-thirds of the articles they may identify in a literature search. Topics covered include: mammals, birds, reptiles, amphibians, behavior, physiology, ecosystems, habitat and distribution, foods and feeding, captive-animal care, and endangered species.

NISC (The National Information Services Corporation) sells individual and institutional licenses to a growing number of online database products. They also offer *Walker's Mammals of the World*, which is a primary reference for zoo personnel. I recently spoke to NISC about the possibility of contributing citations to their database. (The Library maintains an *Endnote* file of all staff articles and many of the researchers and curators have personal *Endnote/ProCite* files on a variety of species.) NISC was extremely helpful in facilitating this plan and it is one we are currently pursuing. *BiblioLine* has future plans to link their citations to any full-text journal subscriptions we may have through EBSCO and *BioOne*.

BioOne *www.bioone.org*

BioOne is a growing collection of full-text Made possible by a cooperative alliance of libra publishers and academia, they offer cost-effec most 60 titles (including *American Zoologist,* *BioScience, The Condor, Copeia, Evolution, Jo of Zoo and Wildlife Medicine,* and *The Wilson* titles for the curatorial staff who lack the Uni searchers.

Primate Lit *www.primate.wisc.edu*

Devoted solely to the primatology literature site has documents found nowhere else: journal tations and fact sheets (complete with reference and Wisconsin Regional Primate Centers are re

Herplit *www.herplit.com/herplit*

Breck Bartholomew does for reptiles and Centers do for monkeys and apes. Bibliomani herpetological material. The *Herplit* Databa 50,000 citations dating from 1586 to the presen

PubMed *www.ncbi.nlm.nih.gov/entrez/query.fc* and ***Agricola*** *www.nalusda.gov/ag98/*

These extremely powerful databases are im for both vets and researchers.

Biosis Previews *www.biosis.org/products_serv*

Although the library does not subscribe to th being regularly used by staff with University p

IMPORTANT WEB RES

The Library Portal establishes hyperlinks to maximize budget dollars. Most links are to st sites, ensuring that the data are reliable. Major s ing substantial full-text information are preferred. A geographic bias is employed favoring California, the southwest, and areas important to our research interests. The survey requested feedback on the Animal, Veterinary Medicine, and Conservation sections of the Portal. Brief descriptions of these key resources are provided below.

ANIMAL INFORMATION

This hierarchy provides links to animal fact sheets; quality Web resources on mammals, birds, reptiles, amphibians and invertebrates; animal husbandry and enrichment; photos, videos and vocalizations; taxonomic resources; regulations and guidelines and our veterinary medicine page.

Society Fact Sheets

Library staff has created a number of unique information resources. One of the most important is a series of fact sheets based on key animals in the Society's collection. The California Condor, Giant Panda, Big Horn Sheep, and Chacoan Peccary are representatives of important conservation projects. Hippos, Polar Bears, Pygmy Chimpanzees, Okapi, and Forest Buffalo are key species in major exhibits. Each fact sheet has a standardized organization, for easy navigation. The following topics are addressed in the same sequence: Taxonomy, Nomenclature & Phylogeny, Distribution & Habitat, Physical Characteristics, Behavior & Ecology, Diet and Feeding, Reproduction and Development, Pathology and Diseases, Captive Management, and Conservation Status. The information is concisely presented in bulleted format and meticulously researched and annotated using the library's extensive resources. All fact sheets are submitted for curatorial review, and complete bibliographies flesh out the brief annotations. The pages are hyperlinked throughout to information on exhibits, current population data, and other important online resources (e.g., the condor sheet links to the AZA Action Plan, the Los Angeles Zoo, the Peregrine Fund, California Fish & Wildlife and USFWS).

Regulations and Guidelines

Links to animal and wildlife laws, regulations, and guidelines are collected on one page for quick access. Representative of the links found there are: APHIS–USDA Rules and Notices; CALIFORNIA FISH & GAME–Legal Affairs; IACUC–Institutional Animal Care and Use Committee; IATA–Live Animal Transportation; USFW–Permits; Guidelines for the Capture, Handling and Care of Mammals–American Society of Mammalogists; International Animal Health Code-2001–OIE.

Mammalian Species
www.science.smith.edu/departments/Biology/V

This series of species accounts was begun in of Mammalogists and now numbers almost 70 count written by an expert in the field. They su tion (at the time it was written) on the species. (biology, distribution, fossil history, genetics, a ecology, and conservation. The accounts vary Some 631 are available in downloadable PDF fi of new mammals is random–based on the intere is an incomplete resource.

Mammal Species of the World *nmnhgoph.si.e*

This resource is the online version of the imp published in 1993 plus subsequent updates. It w national experts under the auspices of the Amer and is considered an authoritative checklist of al

ISIS *www.isis.org*

Zoos have always maintained the pedigree a in their charge. These studbooks have been ava since 1974 and are currently searchable online contributing institutions. Approximately one-t one-third from Europe and one-third from other major overhaul with the addition of modules on bandry, and environmental monitoring. The n ZIMS (Zoological Information Management S velopment is expected to go out in early 2003.

African Mammals Databank *gorilla.bio.uniro*

The Institute of Applied Ecology (IEA), in can institutions has produced this GIS-based da conservation of all the big and medium-sized collected on 281 species (excluding elephant a rhinos (*Diceros bicornis* and *Ceratotherium si* security reasons and the elephant (*Loxodonta af* ephant Specialist Group of the SSC/IUCN.

IUCN Cat Specialist Group *lynx.uio.no/catfolk/*

Developed by Peter Jackson of the IUCN/SSC this is an extraordinary resource–both thorough and authoritative. It is searchable by species or geography.

Ultimate Ungulate *www.ultimateungulate.com*

Maintained by Brent Huffman, this site has very basic full-text information. Emphasis is on the Artiodactyla and Perissodactyla.

Primate Lit Fact Sheets *www.primate.wisc.edu/pin/factsheets/index.html*

This site offers an extensive set of primate fact sheets complete with references and photos. An additional 1,000 fact sheets have been identified from various Web sites around the world and links to these are provided at www.primate.wisc.edu/pin/factsheets/links.html.

Chimpanzee Cultures *chimp.st-and.ac.uk/cultures3/default.htm*

This unique site describes behavioral variations among groups of chimpanzees across study sites in Africa. Searches can be done on species or behavior-type. There are some full-text articles.

International Rhino Foundation *www.rhinos-irf.org/index.htm*

A very good up-to-date resource on all five species of rhinos and their conservation. The technical programs section has tools for professionals. The Rhino Research Center has a searchable bibliography of 7,000 references to relevant books and journal articles.

EMBL Reptile Database *www.embl-heidelberg.de/~uetz/LivingReptiles.html*

This German site describes more than 8,000 species. It is an incredible taxonomic resource (reptile taxonomy is extremely complicated) and is overseen by Peter Uetz. It includes photographs, references and links.

VETERINARY MEDICINE INFORMATION

This section provides links to the major vet med portals (*NetVet*, *Bio Sites*, Martindales' Virtual Veterinary Center, *International Veterinary Information Service* (*IVIS*)); the veterinary literature (*PubMed*, *Agricola* and *WildPro*); diagnostic aids (*Consultant* and Merck); anesthesia and restra tabases; full-text specialty resources and allied hea and CDC). Singled out for special mention are:

Wildlife Information Network–WildPro *institutic*

WildPro, a product from the UK is acknowle thologists as being quite valuable. The goal of comprehensive encyclopaedia of wildlife healt UK fauna, it is fully referenced and provides de identification; natural history; catching, handli and housing; rehabilitation and release; commo It also includes a database of wildlife legislati team of veterinary surgeons are responsible for

The Library has a corporate subscription whi cludes a full year's unlimited access to the *Wild* CD-ROMs for field access. WildPro's Electroni hyperlinked full-text publications: *Wildlife: Firs ease Investigation and Management*; *USGS Fiel Waterfowl: Health & Management*; *Managing f the U.S.* and *Managing Foot and Mouth Disease Guidelines*; *British Mammals Factsheets*; and *En ual*. The taxonomic structure for the *SPECIES* s all mammals, birds, reptiles, and amphibia to ge tion varies greatly in degree of detail. There are but their *Diceros bicornis* prototype is impressi

Currently, link pages have been established f fected by Foot-and-Mouth Disease, and links f Virus are in the works.

Consultant *www.vet.cornell.edu/consultant/co*

A useful diagnostic tool from Maurice White of Veterinary Medicine. Search by diagnosis or

Merck Veterinary Manual *www.merckvetman*

The most comprehensive electronic animal c

Comparative Placentation *medicine.ucsd.edu/c*

Created by Kurt Benirschke of UCSD, this si ans and veterinary pathologists. It enables the ex large number of mammalian placentas.

Heard's Zoological Restraint & Anesthesia *www.ivis.org/special_books/Heard/toc.asp*

McKenzie's Capture & Care Manual *wildlifedecisionsupport.com*

Both of these sites are outstanding veterinary resources and require registration. The International Veterinary Information Service (IVIS) Website provides free access to original, up-to-date publications organized in electronic books as well as proceedings, short courses, image collections, and much more.

CONSERVATION INFORMATION

This hierarchy has links to Society conservation projects, endangered and threatened species lists from the IUCN, USFWS, and CITES; key information resources (NBII, USGS, World Species List); conservation organizations (AZA, IUCN, WWF, Conservation International); important California conservation information resources (California Dept. of Fish and Game); maps and GIS (World Atlas of Biodiversity, African Mammals Databank). The following were visited most frequently by staff:

AZA *aza.org*

The American Zoo and Aquarium Association (AZA) Web site is a key information resource for all zoo staff. There is a great deal of full-text information, but some is restricted to members only. Like the IUCN's Species Survival Commission, AZA coordinates and supports Species Survival Plans. There are more than 100 programs involving 159 individual species managed by experts from North American zoos. They develop master plans that deal with captive breeding (designed to achieve maximum biodiversity), reintroduction and in situ conservation projects. They also publish husbandry manuals (diet care and standard management practices) and contribute to studbook compilations.

One of AZA's most important functions is to facilitate member communication and information sharing by hosting a number of listservs. Zoo researchers subscribe to AZA Conservation and Science, various Animal TAGs and SSPs and the AZA "green" list.

IUCN/SSC *www.iucn.org/themes/ssc/index.htm*

The IUCN (formerly the International Union for Conservation of Nature and Natural Resources) is now known as the World Conservation Union. One of their six Commissions, the Species Survival Commission (SSC) is an inter-

national network of wildlife experts that provid
vice on plant and animal conservation. Consist
they produce the *IUCN Red List of Threatene*
plans for species conservation, as well as a large
establish policy guidelines, organize workshops,
projects. The online database is searchable by co
classifies species according to their extinction ri
formation on species range, population trends, m
conservation measures. An important collection
the SSCs, and our researchers subscribe to one or
of interest (Alien-L, for example, deals with inv

CBSG *www.cbsg.org*

Part of the IUCN, the Captive Breeding Speci
Ulie Seal and is one of the Species Survival Co
Groups. Their reports (Conservation Assessmen
and Population Habitat Viability Assessments–F
web site and are critical for Zoos developing con
also features a global zoo directory to facilitate r

ZOOLOGICAL SOCIETY NEWS AND CU

Two additional Portal resources have proven
The News section is updated every two weeks on
cludes abstracts from major news Web sites, sci
Animal and Plant Health Information Service, a
Service. This current awareness service was req
ing strategic planning sessions in 1999, and has e
tocopies of articles on a monthly basis to a mo
version. The articles posted are a mix of zoo, c
tourism and business trends. It enables staff (f
country) to remain current on the latest develop

For example, synopses of the following arti
2002: *Three Species of Elephants in Africa, Con
sis, A New Mission at Bronx Zoo, WWF: Asian R
bird Decline Linked to Acid Rain, West Nile U
Rights, Alien Invaders vs Conservationists, In
covered, Managing Earth–Even Eden is Engin
Third Birthday, Cloning the Woolly Mammoth,*

mals, Lionesses Go for Dark Manes, Tumucumaque National Park–Brazil, Johannesburg World Summit, New Conservation Online Resource.

Newspapers monitored include: *The New York Times* (particularly the Tuesday Science edition), the *San Diego Union Tribune*, and the *Los Angeles Times*. The journals *Nature*, *Science*, and *Conservation Biology* are regularly monitored. The following News Web Sites are valuable resources: *Eurekalert*, www.eurekalert.org; *BBC Science & Technology News,* news.bbc.co.uk/1/hi/sci/tech; *USDA's Animal and Plant Health Inspection Service,* www.aphis.usda.gov/lpa/press/press.html; Academic Press and Science Magazine's *inScight*, www.academicpress.com/insight/archive.htm; *SciDev.Net* (articles from *Science & Nature*), www.scidev.net; *Yahoo News–Endangered Species,* news.yahoo.com; *Environmental News Network,* www.enn.com; USFWS News, fws.gov/newsreleases; *Planet Ark,* www.planetark.org. Peter Dickenson's *ZooNewsDigest* and Allen Salzberg's *HerpDigest* are useful news-oriented listservs.

In addition, proposed rules, regulations, permit applications and announcements on endangered species from the *Federal Register* are posted to the Intranet. The *Federal Register* listserv on endangered species provides easy access to this information. The service is described at: www.epa.gov/fedrgstr/subscribe.htm.

Table-of-contents links were acknowledged as being regularly used. Offerings are constantly expanding as more and more journals come online, and links are provided to scientific and natural history journals, newspapers, business journals, and major popular magazines.

INTERLIBRARY LOAN

In an era in which full-text online resources provide instant gratification for many, staff still relies on the library to facilitate interlibrary loans. Requests are quite challenging, as library users have become very resourceful in finding the "easy ones" on their own. The articles they do request are frequently produced by small nonprofit societies whose publications have limited circulation. Many are from foreign countries and quite a few are extremely old with jumbled bibliographic citations.

The following e-mail was recently received from a CRES researcher at the Wild Animal Park:

> I am wondering if it is at all possible to obtain a copy of a paper in an Indonesian journal. The name of the journal is "HAYATI Jurnal Biosains" (ISSN: 0854-8587) and the reference is Vol. 8, No. 4, December 2001. I don't know the page numbers but the title (in English) is "The Purity Test

of Bali Cattle by Haemoglobin Analysis
Method." Authors are Ronny Rachman
Maman Armita.

Although five libraries in OCLC claimed to
to supply the article. After locating the journal o
identified and a series of e-mails began. This ki
author:

I got your email from HAYATI Journal.
article using email. Unfortunately the ar
However I can answer any question relate
ther explanation. You can contact me at m
ulty of Animal Science, Bogor Agricult
INDONESIA

We forwarded the information to the researche
rectly. She reported that he was very helpful in

Frequently it is a matter of instructing an i
trieve a document themselves. In answer to a
ceedings of the Cheetah SSP Workshop (Marc
individual on how to download the pdf docume
Posting documents online in pdf format is a gro
ganizations.

Another trend is for listservs to serve as ILL s
Medical Librarians. The following three recent
the San Diego Zoo Library. They illustrate the e
resources and the continued value of older mat

QUESTION 1: I'm working on com
attacks on humans. As you can imagine, t
down, and I would greatly appreciate you
sions, sources, etc. I've searched *Current*
Worldwide (1970?-present) *Zoological*
WorldCatalog.

SDZOO: The San Diego Zoo can sup
phant-inflicted injuries" by K. Benirschke a
ephant attacks through 1995. This docume
numerous sources (wildlife centers, unpub

The document was sent to the requestor and fou
unaware of its existence.

QUESTION 2: One of our professors needs an illustration or written description of the anatomy of the bones in the forelimb of a polar bear (Ursus maritimus). If anyone has any sources that might provide this information, we would be most appreciative. Thank you.

SDZOO: Anatomy questions are always difficult! Shoulder height for adults: 130-160 cm (Mammalian Species #145 pp. 1-7) I can fax you a drawing of an Ursus arctos (brown bear) skeleton from Ian Stirling's Bears; Magestic Creatures of the Wild, 1993. (The polar bear lacks a shoulder hump and has a longer neck and smaller head than other ursids.) I can also supply drawings/measurements of paws and claws from CITES Identification Manual vol. 1 Mammalia (Convention on International Trade in Endangered Species of Wild Fauna and Flora–United Nations Environment Program).

RESPONSE: The fax of the drawings came through just fine. P.S. I definitely agree about anatomy, especially for non-domestic species. At every opportunity I've tried to encourage our now retired gross anatomy professors (Dr. Howard Evans and Wolfgang Sack) to compile anatomical sources for exotic species. The knowledge of these individuals needs to be saved before everything goes molecular.

QUESTION 3: I'm stumped by this one, and hope one of you will be able to help. Our vet is writing a grant proposal for chronic wasting diseases in deer and elk and needs normal blood values for these animals. He thinks they would have been published in the 1960s in a monograph (rather than an article) when medical science was just beginning to be able to do this type of blood work. All we have found to date is white blood cell counts, not differential counts. If anyone has any suggestions as to where this historic data might be published, I would be very grateful!

SDZOO: The information you are seeking may be in the *ISIS Physiological Data Reference Values*, 1995. They have differential data by species and breakdowns by sex and age. The information is also provided with ISIS' *MedARKS* software. We have the 1961 Biological Handbook *Blood and other Bodily Fluids* and the *Biological Data Book* vol III, 1974 but they have very little information on deer and elk.

VIRTUAL REFERENCE

Reference service is now facilitated by e-mail and fax. This typical request arrived last month from CRES's senior scientist in Behavior:

>>> Scientist 09/18/02 03:04PM>>>

In the last ZOONOOZ, Robin McLaughlin mentioned that there were 445 Rothschild's giraffe in the wild, and after a series of email inquiries, I found out that the number con hoofstock published in 1992 by IUCN, on a loss as to how to find the paper, so, lu tance!!! Can you track down the referenc

>>> Librarian 09/19/02 09:44AM >>>

I wasn't able to find the paper–my gu number might be wrong. I did find some mc giraffe (Antelope Survey Update, Number 5 fax it to you.

Another request came after the new taxonc firmed for Gorillas in 2000. Previously consid now divided into two species and five subspec CRES was attempting to locate photos of the across one that he found puzzling. He sent the

I found an image said to be Gorilla goril this web site. It sure looks a lot like the What do you think?

It was Albert! By locating the same photo photograph was correctly identified as Gorilla

Some time ago a discussion concerning the ductions into Mexico led to an inquiry about le mental contaminant in both the U.S. and Mex provide an announcement from the *Federal Reg* titled: Migratory Bird Hunting; Temporary App for Hunting Waterfowl and Coots During the 2 ies were described in detail and a bibliography tion appears in May 10, 2002 (Volume 67, Num in *Federal Register* studies are extremely curre ing resource for all types of endangered species

SURVEY FOLLOW

Surveys are important communication tools ar assessing a library's strengths and weaknesses, learn even more by following up on the informat veterinarians provided a cryptic answer to the su

professional journals would you like the Library to subscribe to? When asked for clarification on what "TNTC" stood for, the following exchange occurred:

>>> veterinarian >>>

Oops, sorry. It is laboratory vernacular for "Too Numerous To Count."

>>> librarian>>>

So I'm dealing with a "wise guy"! Seriously . . . What can I get for you? Don't you have everything you need now that you are under the Davis umbrella?

>>> veterinarian>>>

I do not have access to any of Davis' libraries is complaint number one. Complaint number two is that our library does not have a list or reference for what books are in there and how to find them. Complaint number three is that I have no idea what resources are available through your department. Complaint number four is that my front left tire is low on pressure and I've got to get it fixed! My only gratitude is your interest and kindness. Thanks.

>>> librarian >>>

WOW! I think I can answer all of your complaints:
Complaint #1: As an adjunct professor you have access to the resources of the UC-Davis Library. The head of Vet Services can help you set up your web browser for access. Give him a call.
Complaint #2: Although we do not have a university-quality system you can search for our books and journals. From the main menu of the Library Intranet, click on Animals; Then click on Books (search by keyword, author, title or subject); click on Journals for a list of what we have. Please call me when you want to do a search and I'll walk you through it. Also click on Veterinary Medicine when you get a chance. I think you'll find a lot of useful links.
Complaint #3: Hopefully you will be able to find EVERYTHING you need in life through the Library Intranet (Except the air for your left front tire) AND . . . we always treat cute guys with interest and kindness!

Our survey very clearly pointed out that:

1. Communication and a sense of humor are as important as ever!
2. Technology is still a problem for many of our users. They do not yet have an acceptable comfort level with computers or online systems.
3. Everyone is pressed for time. No one has time for bibliographic instruction, but they would appreciate an online tutorial to refer to.

4. The Web, e-mail, and listservs, are havin
   way individuals seek information. Our re
   this new environment–to find a "niche."
5. "If you build it" they won't necessarily c
   out how to do a much better job of promo
   We need to turn up our "ROAR."

APPENDIX

ZOOLOGICAL SOCIETY OF SAN DIEGO LI

Your Department: Curators Veterinary Services

Which libraries or information services do you use: (Cho

Zoo Hospital Library
frequently occasionally rarely never

WAP Hospital Library
frequently occasionally rarely never

Department Library (please specify which one)
frequently occasionally rarely never

Main Zoo Library
Physically frequently occasionally rarely
Online frequently occasionally rarely

UCSD Scripps
Physically frequently occasionally rarely
Online frequently occasionally rarely

UCSD Biomed
Physically frequently occasionally rarely
Online frequently occasionally rarely

SDSU
Physically frequently occasionally rarely
Online frequently occasionally rarely

Another Library (please identify)

What Library privileges do you enjoy at other libraries?
conditions of eligibility, e.g., UCD adjunct professorship/

e-mail
Online access to full-text databases
Direct borrowing
Inter-Library Loan

On average, how often do you use the Library OR the Library's Information Intranet?

- Once a week or more
- Once a month
- A couple of times a year
- Never

Please evaluate the following aspects of the Library (**Add a yes or no answer** after each statement):

The Society's Collection

Books available in the Library are adequate for my research needs.
Journals available in the Library are adequate for my research needs.
Online services/databases are adequate for my research needs.

Reference Services

Reference collection is (good, adequate, poor, can't rate).
Reference/Research Assistance is (good, adequate, poor, can't rate).
Turn-around on information requests is prompt.

Library Information Intranet (10.1.31.4 or by password through zoo.intranet.org) Answer yes or no.

I'm familiar with this resource.
It is well organized and easy to navigate.
It is a good access tool (important hyperlinks to outside resources).
I find the following to be of value:

- Table of Contents
- News
- Federal Register Announcements
- New book notification
- Animal Fact Sheets
- Zoo History Time Line
- Wildlife Review/Biblioline (index to journal articles)
- BioOne (Electronic full-text resource)
- Other (please identify)

Library Online Book Catalog (via 10.1.31.4 or library.sandiegozoo.org)

It is always available online when needed.
It is easy to use.
I have more luck finding books by ignoring the catalog and going directly to the shelves.

Online Databases and Print Journals

Biblioline/NISC is a valuable resource.
BioOne is a good full-text resource.
WildPro is a useful resource.
The Intranet list of journals is helpful.

Interlibrary Loan/Document Delivery

Inter-Library Loan requests are easy to make.
Percentage of loan requests that have been filled?

APPENDIX (continue

Circulation (yes or no)

The Library's circulation services and policies are compa
The renewal process is easy.

Which electronic databases do you find most useful? (Pl

- Agricola
- Biblioline/NISC
- BioOne
- Biosis Previews
- Medline
- Primate Lit
- Herp Lit
- WildPro
- Zoological Record
- Ingenta (popular info)
- Magazine Portal
- Other (please identify)

What professional journals do you personally subscribe t

What additional professional journals would you like the L

Do you have a personal collection of photocopied articles

What 3 books do you refer to most frequently?

Are there additional Web sites that you feel the Library Intra

What Current Awareness tools do you use? (Add yes or

- Library Intranet News Service
- Library Intranet Federal Register Announcements
- Library Intranet Table of Contents
- Browse recent issues of print journals
- Outside Alerting Service (please identify)
- Listservs (please identify)

What barriers do you experience when seeking informati

- Difficulty securing needed resources (price/availabili
- Information retrieval difficulties (technology problems
- Not enough time
- Inconvenience (physical location of our library, library h

If the online version of a journal is available, should the L

Library Instruction/Training (Please add the words 'works
to indicate your training preference)

I would like instruction on using the Internet/Web.
I would like instruction on using the Library's Intranet Info
I would like instruction on searching online databases.

The Society Library has had a positive impact on my wor

What areas of the library could be improved?

Thank you very much for completing this survey, your inp

Botanical Information: Resources and User Needs

Susan Fraser

SUMMARY. Botanical libraries that once specialized only in the subject of botany now take a more interdisciplinary approach in order to meet the needs of a wide variety of patrons. Information regarding the plant sciences is critical for economic botanists, plant taxonomists, historians, anthropologists, phytochemists, geneticists, biologists, Ph.D. students and graduate fellows. These specializations reflect the academic interests of the researchers who make up the broad user community of the Mertz Library of the New York Botanical Garden. While these scientists agree that powerful electronic search tools have revolutionized their research, hands-on use of the library, scouring the published literature for primary references, and the ability to view journals onsite remain high priorities in their information seeking. *[Article copies available for a fee from The Haworth Document Delivery Service: 1-800-HAWORTH. E-mail address: <docdelivery@haworthpress.com> Website: <http://www.HaworthPress.com> © 2001 by The Haworth Press, Inc. All rights reserved.]*

KEYWORDS. Botanists, botanical information, plant names, plant science, botanical literature

---

Susan Fraser, MLS, is Head of Information Services and NYBG Archivist, LuEsther T. Mertz Library, The New York Botanical Garden, Bronx, NY 10458-5126 (E-mail: sfraser@nybg.org).

[Haworth co-indexing entry note]: "Botanical Information: Resources and User Needs." Fraser, Susan. Co-published simultaneously in *Science & Technology Libraries* (The Haworth Information Press, an imprint of The Haworth Press, Inc.) Vol. 21, No. 3/4, 2001, pp. 121-129; and: *Information and the Professional Scientist and Engineer* (ed: Virginia Baldwin, and Julie Hallmark) The Haworth Information Press, an imprint of The Haworth Press, Inc., 2001, pp. 121-129. Single or multiple copies of this article are available for a fee from The Haworth Document Delivery Service [1-800-HAWORTH, 9:00 a.m. - 5:00 p.m. (EST). E-mail address: docdelivery@haworthpress.com].

http://www.haworthpress.com/store/product.asp?sku=J122

10.1300/J122v21n03_08

INTRODUCTIO

Technology has changed the way we look at t search for information about it. Librarians pro whenever possible, free access to multiple dat The librarian has become an information man and organizing information in a variety of me skills remain essential, the librarian now embra locate, manipulate, filter, and present the infor

Access to the published literature has been ma enter a few keywords into the right database to re sults. In a matter of seconds one can locate a ti throughout the country by the use of bibliograph line Computer Library Center) and RLIN (the I Network). This paper describes some of the sp tronic and printed, used by the broad client base

INFORMATION-SEEKING BEHAV

Whether located within a botanic garden, an or a university, botanical libraries generally ser clude plant taxonomists, systematists, horticult uate students, and the general public. These related to the plant sciences in order to meet th

An enormous amount of botanical informati thus surfing the World Wide Web (WWW) is o information. A recent study of information-ga faculty indicate that researchers in the field of b tronic resources as much as 78% of the time.[1] they verified the information through another s Many botanists use *Google*[2] as a search tool and in locating obscure information as well as for ot language translation capabilities. However, scie alike must guard against the abundance of non-p be posted on the WWW.

While many botanists create and post web p research, taxonomic descriptions must still be format to be valid. Since most of the cited literat lished in serials, scientists in this field value the noncurrent journals in their libraries.[3] As in oth

to e-journals is a priority with plant scientists who, working in remote locations in the field as far away as the Peruvian Amazon, may retrieve published articles and reliable, current scientific data in a timely manner. As these scientists become more dependent on remote retrieval systems, they rely more heavily on JSTOR, the scholarly electronic journal archive that contains almost 30 journal titles in Ecology and Botany.[4]

The botanical community has its own broad network of information providers and collaborators and, as in other disciplines, the Internet and the use of e-mail facilitate communication.[5] Most botanists queried for this article indicated that they maintain their own databases, making an effort to standardize fields to conform to the guidelines being set forth by the International Working Group on Taxonomic Databases for the Plant Sciences (TDWG).[6] Thus, information can be more easily shared within the colleagues and research associates.

Given the interdisciplinary nature of the science of botany and the diverse information needs of the botanical community served, information managers must keep abreast of certain legal, political, medical and copyright issues that impact the use of the information. As new fields of study such as genomics and molecular systematics emerge in the study of plants, libraries are pressured to expand their ability to access these new disciplines.[7] In addition, they must be equipped to interpret and instruct users in the use of chemical databases, online atlases and gazetteers and other subject areas beyond the initial scope of the library. Librarians and other information specialists respond by providing bibliographic instruction to databases and other search tools and engines.

ELECTRONIC RESOURCES

Botanists have identified these electronic resources, once only available in printed format, as being indispensable tools in their research. Today, field botanists and collectors take laptops, palm pilots, camcorders, digital cameras, and a host of other technologies into the field to document their collecting methods and research. They can then link photographic documentation and research data to the plant specimens collected. Geographic Information Systems (GIS) have become indispensable for mapping out collecting areas. By inputting descriptive attributes on map points, the botanist can pinpoint geographic areas where species are most likely to occur. Digitized georeferenced maps can be located on the Royal Botanic Gardens, Kew Website.[8] Online gazetteers are used frequently by botanists and curatorial staffs of herbariums to verify places, names and collecting localities. GEOnet[9] and the *Getty Thesaurus of Geographic Names*[10] are considered favorites, but there other good sites especially for locating and verifying more specific locations such as Brazilian municipios (counties) through the *IBGE Cidade* and herbarium study, the botanist is also emplo ory, molecular techniques including DNA se morphological variation, the sectioning of org light and electron microscopy. Outstanding elec searcher locate relevant periodical titles in m BIOSIS, or *Biological Abstracts*,[12] AGRICOL ricultural but includes soil science and entomc States Department of Agriculture (USDA)[13] anc areas of forestry, genetics and biotechnology. H resources. Medline or PubMed[15] is available th Medicine for access to medical journals but als ences, and OCLC FirstSearch[16] includes *Dis* and the Union list.

Powerful searchable databases include onlir Mertz Library's CATALPA,[17] which has been se since 1994 and is recognized as an international botanists to share and access data essential for th Missouri Botanic Garden's vast nomenclature d International Legume Database & Information S a detailed directory of public herbaria of the worl members associated with them along with stanc *Index to American Botanical Literature*,[21] a cor the Americas.

Given the historic nature of the science of bot the older literature as well as currently publishec resource used for verifying nomenclature is the *I* (IPNI)[22] which combines the resources of *Ind* Royal Botanic Gardens, Kew, the *Gray Card Ind* Herbaria, and the *Australian Plant Names Inde* Herbarium. The *IPNI* is a database of the flower bibliographical details for citations relating to th

The *International Code of Botanical Nomen* *Genericorum*[24] are tools used by botanists to ve tional information about plant and plant names Department of Agriculture's Natural Resources National *PLANTS* database.[25]

The International Union for Conservation of produces *The IUCN RedList of Threatened Spe* information for plant collectors, students, and h

chemists and students doing laboratory research and studying plant compounds use Chapman and Hall *CRC Chemical Dictionaries on CD-ROM.*

Another invaluable resource is the subscription-based online resource, *Plant Information Online*[27] from the Anderson Horticultural Library at the University of Minnesota, a user-friendly database for sources of seeds and plants, as well as book and journal citations. The *Royal Horticultural Society Plant Finder Reference Library* online, which includes the *RHS Plant Finder 1999-2000*, is useful for finding cultivar names; and the *Dictionary of Common Names* enables users to check both Latin and common plant names.[28]

PRINTED RESOURCES

Scientists agree that technology, while considered essential, cannot replace the researcher's ability to browse the current literature onsite in the library. Botanists looking for type descriptions or locality information for a particular collection must often check the original publication. The pleasure of finding an obscure tidbit of information in an unindexed foreign newsletter and the ability to leaf through the monograph or journal collections and review the current literature remains a primary means of seeking information.

Specialty societies like the American Fern Society and the Cycad Society, to name but two, provide a venue for professionals and amateurs alike to access specific information about particular plant families and to keep abreast of current discoveries in their subject area. While many societies maintain Web sites, some do not. Small society newsletters are often overlooked in the commercially available electronic indexes; therefore it is crucial to the scientists whose research is focused on a particular area of study to review these unindexed publications. In addition to newsletters, conference proceedings, bulletins and other grey literature fall into this category.

Other printed resources used regularly by scientists and students include D.J. Mabberley's *The Plant Book*, 2nd edition, a comprehensive dictionary and an essential tool for information on plant names and uses. Bibliographies such as *Taxonomic Literature 2*, a comprehensive listing for all authors published before 1940, compiled by Frans Stafleu and Richard Cowan, is a standard work and essential for the systematist or botanical librarian. It provides the location of botanical collections and accurate bibliographic and biographical information, along with some descriptive commentary. *Botanico-Periodicum-Huntianum*, or BPH as it is fondly known, provides full citation information for abbreviated botanical journal and serial titles. Prepared under the auspices of the Royal Horticultural Society of London, *Index Londinensis* is an index to figures of flowering

plants, ferns and fern allies scattered throughout t
and is an indispensable source for picture searchi

Individual journals and monographic and *Intermountain Flora*, standard flora of the *Brasiliensis*, a standard for Brazilian flora, *Flor* *mala*, published by The Field Museum in Chicag botanists.

UNPUBLISHED LITER

Access to unpublished literature and other pr way into databases. The Research Libraries Grou Cultural Materials Initiative[30] has resulted in dat als such as manuscripts, correspondence, phot cultural artifacts. Contributions to these indexe freestanding museums including natural history dens. The archival collections held at The New Y late to the history of botany and botanical explor the RLG's Archival Resources, and they can be well as on the Garden's Web site.[31]

COLLABORATIO

Collegial arrangements such as that with the ticultural Libraries (CBHL),[32] organized in 196 foster the advancement of botanical and horticul resource sharing, and its European counterpart, reliable and steadfast librarians who serve as an ber libraries. A CBHL listserv, available to me change of botanical or library related informa collection development policies, database desig discuss professional development issues; or eve tions or a hard-to-find citations. The query is th ship via e-mail and through a network of libraria the question is researched and answered.

The organization shares duplicate title lists which serves as an avenue for alerting members that can then be shared with the library clientele portant element for libraries to better support th serve.

INSTITUTES AND PROGRAMS

The LuEsther T. Mertz Library of the New York Botanical Garden is one of the world's largest and most active botanical/horticultural libraries. The Library collects, catalogs, conserves and makes available to all users the published literature and the unpublished documentary record in the subject areas of systematic, floristic and economic botany, plant ecology, horticulture and gardening, landscape design, garden history and botanical and horticultural bibliography and biography.

The principal clientele of the Mertz Library is the *research staff*, primarily from the Institute of Systematic Botany, the Institute of Economic Botany, The Cullman Program for Molecular Systematic Studies, and the Graduate Studies Program;[34] the *curatorial staff* of the Herbarium and Horticulture departments; and *students* in the School of Professional Horticulture. The Library also supports the Continuing Education Department and the general public.

The Institute of Systematic Botany (ISB) supports systematic research on the taxonomy and conservation of vascular and cryptogamic plants. Garden scientists collect and preserve plant specimens throughout the world, particularly in South and Central America and the American West, documenting biological diversity and employing conservation strategies to study and preserve the world's ecosystems. The Garden's Herbarium contains over 6,500,000 specimens; and the Garden's Virtual Herbarium[35] an online resource containing high resolution scans of herbarium specimens, is now available for over 84,000 vascular plant type specimens. The Virtual Herbarium has served as a model project, making the unique type specimen collection available to remote users. The Virtual Herbarium enables botanists across the globe to study a type specimen online, often eliminating the need to borrow the specimen on loan, as well as responding to the delicate and tenuous balance between preservation and access.

Scientists in the Institute for Economic Botany (IEB) study the Interaction of Plants and Mankind. Their research takes an interdisciplinary approach to applied research in the biological and social sciences. Staff in the IEB is currently involved in a broad-scale survey of U.S. plant biodiversity to identify phytochemicals with potential use in the pharmaceutical industry. Garden scientists collect plant material and analyze its chemical properties.

The Cullman Program for Molecular Systematic Studies involves the study of plant molecular biology, specifically the evolution of genes and their functions. These scientists study and document the earth's biodiversity by researching and studying plant genes, evolutionary history, classification, biogeography, phytochemistry and ethnobotanical uses. Sequencing the plants' DNA provides information on how organisms are related. Laboratory studies are conducted for both chemosystematic research and for biodivers basic work consists of identifying and understand ary plant compounds.

The Garden has long had a tradition of trainin ration with universities includes Columbia Uni mental Conservation (CERC), Yale's School of Studies, Cornell University, New York Universi New York. The Library provides service to stu well as to students with fellowships from other cc ate program includes 44 students from 18 count

Whether documenting the earth's flora, studyi cessing genetic resources or contributing to a glc vation, scientists need access to a multitude of shift towards virtual collections, the librarian ha manager, responsible for the task of getting the p sources they need in a format they can use.

REFERENCES

1. Carlson, Scott. Students and Faculty Members Before Printed Ones, Study Finds. *Chronicle of Highe*

2. *Google* (www.google.com) (http://www.googl cessed November 2002].

3. Bonn, George and Linda C. Smith. 1992. Liter *McGraw Hill Encyclopedia of Science and Technolog*

4. *JSTOR*–The Scholarly Journal Archive. http: vember 2002].

5. Tenopir, Carol and King, Donald W. 2000. *T alities for Scientists, Librarians, and Publishers*. Spec ington D.C.

6. Brummitt, R.K. 2001. *World Geographic Sche tions*. Ed. 2 Published for the International Working G Plant Sciences (TDWG) by the Hunt Institute for Botani

7. Davis, Elisabeth B. and Diane Schmidt. 199 *ture: a Practical Guide*. New York. Marcel Dekker,

8. Royal Botanic Gardens, Kew (http://www.rb November 2002].

9. GEOnet, (http://164.214.2.59/gns/html/index.htm

10. The Getty Thesaurus of Geographic Names tools/vocabulary/tgn/index.html) [Accessed November

11. IBGE Cidades@ http://www1.ibge.gov.br/c November 2002].

12. BIOSIS, (http://www.biosis.org/index.htm) [A

13. United States Department of Agriculture. AGRICOLA (http://www.nal.usda.gov/ag98/) [Accessed Novmeber 2002].

14. CAB abstracts (http://www.cabi.org/) [Accessed November 2002].

15. National Library of Medicine. Medline or PubMed (http://www.ncbi.nlm.nih.gov/entrez/query.fcgi) [Accessed November 2002].

16. OCLC FirstSearch (http://www.oclc.org/firstsearch/databases/) [Accessed December 2002].

17. The New York Botanical Garden. CATALPA. (http://www.nybg.org/bsci/libr/catalog.html) [Accessed December 2002].

18. Missouri Botanical Garden. TROPICOS, (http://www.mobot.org/W3T/Search/vast.html). [Accessed November 2002].

19. The International Legume Database and Information Service (http://www.ildis.org/LegumeWeb/) [Accessed December 2002].

20. The New York Botanical Garden. *Index Herbariorum* (http://wwwnybg.org/bsci/ih/) [Accessed November 2002].

21. The New York Botanical Garden. *Index to American Botanical Literature.* (http://www.nybg.org/bsci/iabl.html) [Accessed December 2002].

22. *The International Plant Names Index* (IPNI) (http://www.ipni.org/) [Accessed December 2002].

23. The *International Code of Botanical Nomenclature* (http://www.bgbm.fu-berlin.de/IAPT/) [Accessed November 2002].

24. *Index Nominum Genericorum* (http://rathbun.si.edu/botany/ing/ingform.cfm) [Accessed November 2002].

25. United States Department of Agriculture's Natural Resources Conservation Service (NRCS) National PLANTS database (http://plants.usda.gov/) [Accessed May 2003].

26. The International Union for Conservation of Nature and Natural Resources *The IUCN RedList of Threatened Species* (http://www.redlist.org) [Accessed November 2002].

27. University of Minnesota. *Plant Information Online* (http://plantinfo.umn.edu/arboretum) [Accessed November 2002].

28. The *Royal Horticultural Society Plant Finder Reference Library* (http://www.rhs.org.uk/rhsplantfinder.asp) [Accessed November 2002].

29. The Research Libraries Group's Archival Resources. (http://www.rlg.org/arr/index.html) [Accessed November 2002].

30. The Research Libraries Group Cultural Materials Initiative. (http://culturalmaterials.rlg.org/cmiprod/workspace.jsp) [Accessed October 2002].

31. The New York Botanical Garden. Archival resources. (http://www.nybg.org/bsci/libr/archives.html) [Accessed December 2002].

32. The Council of Botanical and Horticultural Libraries (CBHL) (http://www2.ville.montreal.qc.ca/jardin/cbhl/cbhl.htm) [Accessed November 2002].

33. European Botanical and Horticultural Libraries *(EBHL)* (http://www.ub.gu.se/Gb/ebhl/home.htm) [Accessed October 2002].

34. The New York Botanical Garden. Science Programs. (http://www.nybg.org/bsci/) [Accessed December 2002].

35. The New York Botanical Garden Virtual Herbarium (http://www.nybg.org/bsci/hcol) [Accessed December 2002].

Distinguishing Engineers from Scientists–The Case for an Engineering Knowledge Community

Thomas E. Pinelli

SUMMARY. This article makes the case for an engineering knowledge community. We begin by discussing the differences between science and technology. We next discuss the similarities and differences between engineers and scientists. Next, we analyze previous research into the information use behaviors of engineers. Finally, using the research results from the NASA/DoD Aerospace Knowledge Diffusion Research Project, we compare and contrast aerospace engineers and scientists as a means of developing similarities and differences between engineers and scientists in terms of their information-seeking behavior. The goal of this article is to demonstrate that engineers are not scientists and that knowledge production and use differ in engineering and science. We believe that the current model used to explain information-seeking behavior assumes no difference between the information-seeking of engineers and scientists. The distinctions between engineering and science, engineers and scientists and the information-seeking behaviors of engineers and scientists have multiple implications for providing information services, knowledge management, and diffusing knowledge. The message to libraries is "know thy customer."

Thomas E. Pinelli, PhD, is Educational Technology and Distance Learning Officer, Mail Stop 400, NASA Langley Research Center, Hampton, VA 23681-2199 (E-mail: t.e.pinelli@larc.nasa.gov).

[Haworth co-indexing entry note]: "Distinguishing Engineers from Scientists–The Case for an Engineering Knowledge Community." Pinelli, Thomas E. Co-published simultaneously in *Science & Technology Libraries* (The Haworth Information Press, an imprint of The Haworth Press, Inc.) Vol. 21, No. 3/4, 2001, pp. 131-163; and: *Information and the Professional Scientist and Engineer* (ed: Virginia Baldwin, and Julie Hallmark) The Haworth Information Press, an imprint of The Haworth Press, Inc., 2001, pp. 131-163.

http://www.haworthpress.com/store/product.asp?sku=J122
10.1300/J122v21n03_09

KEYWORDS. Engineers, scientists, kno
vices, management, knowledge diffusion,

INTRODUCTIO

The relationship between science and technol tinuous process or normal progression from basic plied research (technology) to development assumes that technology grows out of or is depen opment. This "assumed" relationship is the found policy is based and may help to explain the use c entists and engineers." It also helps us to unders and dissemination practices are aimed toward sci technology has its ultimate roots in science perpe tion system that assumes all technology will hav

However, the belief that technological chang tific advances has been challenged in recent ye been increasingly seen as the adaptation of exis response to demand (Langrish, Gibbons, Evans, several years of study that attempted to trace the ence to technology have produced little empiric tionship (Illinois Institute of Technology, 1968 1969). Price (1965), for example, claimed tha are derived immediately from the technology th ence or applied science. Price concluded that sc independently of one another. Technology buil ments and advances in a manner independent o entific frontier and often without any necessit basic science underlying it. Shapley and Roy (1 gression from science to technology does not ex nication between science and technology. R indirectly supported by each other. We posit th ference between science and technology, so too between engineers and scientists and that this " gineers and scientists "seek and use" informati NASA/DoD Aerospace Knowledge Diffusion compare and contrast aerospace engineers and oping "generalizable" similarities and differenc entists in terms of their information-seeking be

Science and Technology

Science is an introverted activity. It studies problems that are usually generated internally by logical discrepancies or inconsistencies or by anomalous observations that cannot be accounted for within the present intellectual framework. Indeed, scientists are said to do their best work when investigating problems of their own selection and in a manner of their own choosing (Bush, 1945; Amabile, 1983; Amabile and Gryskiewicz, 1987). The output of science is knowledge that is regarded by scientists essentially as a free good. The expectation within the scientific community is that knowledge will be made universally available through presentations at conferences and society meetings and publication in scholarly and professional journals.

Technology, on the other hand, is an extroverted activity; it involves a search for workable solutions to problems. When technology finds solutions that are workable and effective, it does not pursue the *why* (Salomon, 1984). Moreover, the output of technology is frequently a process, product, system, or service. Technological knowledge is not easily or completely codified, nor is it freely communicated. Unlike science, the output of technology is *not* made universally available. Technology successfully functions *only* within a larger social environment that provides an effective combination of incentives and complementary inputs into the innovation process. Technology is a process dominated by engineers rather than scientists (Landau and Rosenberg, 1986).

As *social organizations*, science and technology have very different attitudes and values concerning knowledge and its ownership. Generally speaking, the scientific community tends to view knowledge as a *public consumption good*, while engineers (or, more precisely, the firms that employ them) regard it as a *private capital good* (Dasgupta, 1987). Thus, the rules of the two communities concerning the communication (i.e., disclosure) and ownership of knowledge are fundamentally different. Scientists are obligated to disclose their findings and to submit them for critical inspection to other members (i.e., peers) of the scientific community. Hence, the ability of scientists to communicate freely and openly is critical.

Moreover, knowledge production takes place in the context of two very different reward systems. In science, rewards are based on priority of discovery or the *rule of priority*. This rule acts as an incentive for scientific discovery, and serves to promote public disclosure of that discovery. Thus, scientists are compelled to take privately created knowledge and to make that knowledge accessible to the scientific community and the general public. The rule of priority also precludes a second or third place winner because from a societal point of view, there is no value added when a discovery is made a second or third time.

As is often the case, policies based on this arr intended consequences, particularly when com objectives. For example, U.S. science policy sup open communication of federally funded scienc progress. Economic and national security polici strictions and controls on the flow of scientific merous regulations that have fueled pragmatic the open communication of ideas and informatic nical disciplines. This has become a particular tional collaborative ventures.

The attitudes toward and values concerning are different in the technology community. Dis expected in the technology community. In fact cause the reward system is linked to "privately earned from the production of knowledge. Tec sidered proprietary, is afforded patent protection often the subject of industrial espionage. Its use ing to pay an agreed upon price. What is intere technology community is that, although patent vate knowledge public, it does not attempt to value on the knowledge. The worth and risk vately capturable rents from knowledge are left being first to the market often results in financi for a process, product, system, or service to be contrast with science, there can be second and t ogy. In that case, society usually looks for what "best-practice" technique that is often subject sumer as a compromise between price and perfo point, having more than one winner stimula produce multiple and perhaps better products fo

Many researchers have questioned the classi and technology and between scientists and en Many current theories of science and technolo that if researchers make their observations at eit etal level, the distinctions between science an Some theorists of technology studies believe tha termine the technologies that will be developed Law and Callon (1988), for example, argue tha who design societies and social institutions to (1992) argues that the perspective of the researc of where to place activities or actors in science

gues that "the dancing partnership of science and technology [is] a relation between activities oriented to different reference points and groups, rather than a matter of combining different cognitive-technical repertoires" (p. 257). That is, science and technology, scientists and engineers do many of the same activities but in different ways.

The distinction is further clouded when one looks closely at the varieties of actors and organizations that constitute technology. For example, in aerospace some engineers and scientists are working on methods to explore the edge of the universe and others on how to best design an aircraft for passenger comfort. Some deal with very abstract ideas and others with difficult technological, economic, or management issues. Much research that attempts to understand the differences between science and engineering has examined what Constant (1980) termed radical science or technology. That is, much research focuses on changes in paradigms or fundamental ways of thinking about a phenomenon or artifact. For example, Constant (1980) examined the role of presumptive anomalies in technology to understand fundamental changes. His best example is the adoption of the jet engine. Little research focuses on the day-to-day activities of scientists and engineers where science and technology are maintained through routinized activities.

Allen's (1977) study of the transfer of technology and the dissemination of technological information in R&D organizations found little evidence to support the relationship between science and technology as a continuous relationship. Allen concluded that the relationship between science and engineering is best described as a series of interactions that are based on need rather than on a normal progression. According to Allen, the results of science do progress to technology in the sense that some sciences such as physics are more closely connected to technologies such as electronics, but overall, a wide variation exists between science and technology. A direct communication system between science and technology does not exist. The most direct communication between science and engineering takes place through the process of education.

In summarizing the differences between science and technology, Price (1965) made the following 12 points:

1. Science has a cumulating, close-knit structure; that is, new knowledge seems to flow from highly related and rather recent pieces of old knowledge, as displayed in the literature.
2. This property is what distinguishes science from technology and from humanistic scholarship.
3. This property accounts for many known social phenomena in science and also for its surefootedness and high rate of exponential growth.

4. Technology shares with science the sa shows quite complementary social phen tude to the literature.
5. Technology therefore may have a sin structure to that of science, but the stru rather than of the literature.
6. Science and technology therefore have t structures.
7. A direct flow from the research front of or vice versa, occurs only in special a structures are separate.
8. It is probable that research-front technol that part of scientific knowledge that has ambient learning and education, not to r
9. Research-front science is similarly relate logical knowledge of the previous genera search front of the technological state of
10. This reciprocal relationship between sci ing the research front of one and the ac nevertheless sufficient to keep the tw growths within each one's otherwise ind
11. It is naive to regard technology as applie
12. Because of this, one should be aware of a entific research is needed for particular and vice versa. Both accumulations are separate ends (Price, 1965, pp. 557-563)

Allen (1977) also stated that the independent r ogy and the different functions performed by en influence the flow of information in science and nology are ardent consumers of information. Bo quire large quantities of information to perform t is a strong similarity between the information inp entists. However, the difference between engine information processing becomes apparent upon Scientists use information to produce informatio the input and output, both of which are verbal, are one stage is in a form required for the next stage. produce some physical change in the world. Eng transform it, and produce a product that is infor information is no longer in verbal form. Wherea duce information in the form of human languag mation from a verbal (or often, visual or tacit) fc form. Verbal information is produced only as a by-product to document the hardware and other physical products produced.

Allen finds an inherent compatibility between the inputs and outputs of the information-processing system of science. Because both are in verbal formats, the output of one stage is in the format required for the next stage. The problem of supplying information to the scientist is a matter of collecting and organizing these outputs and making them accessible. Since science operates for the most part on the premise of free and open access to information, the problem of collecting outputs is made easier.

In technology, however, there is an inherent incompatibility between inputs and outputs. Since outputs typically differ in form from inputs, they usually cannot serve as inputs for the next stage. Further, the outputs are usually in two parts, one physically encoded and the other verbally encoded. The verbally encoded part does not serve as input for the next stage because it is a by-product of the process and is itself incomplete. Those unacquainted with the development of the hardware or physical product therefore require some human intervention to supplement and interpret the information contained in the documentation. Since technology operates to a large extent on the premise of restricted access to information, the problem of collecting the documentation and obtaining the necessary human intervention becomes difficult.

Allen and others used a somewhat restricted definition of technology in that they assume that it is always a physical product. Engineers in aerospace and in other industries often create systems and products that are verbally encoded, such as management systems and software. These differences do not alter the basic premise that substantial differences exist between the goals of engineers and scientists as they produce the different types of outputs in their daily activities. The connection between science and technology, in aerospace and elsewhere, is tenuous, vague, and sporadic. The processes used in science and technology to produce their respective outputs create parallel and weakly connected systems. A clear recognition of these differences is needed to establish a context for and to understand aerospace knowledge diffusion (i.e., production, transfer, and use).

Engineers and Scientists

For our purposes, we define the essential difference between engineers and scientists based on the primary goal of the output of their work–scientists produce knowledge (facts) and engineers produce designs, products, and processes (artifacts). Engineers and scientists exhibit many other important differences in education, technical discipline, and type of work activities. These differences point to differences in their information-seeking behaviors and information

needs. In this section, we describe many of the d
ences in characteristics and proceeding to differ

Differences between engineers and scientists
either self-classification or the analysis of their
pp. 26-56) describe differences based on analy
education, and self-identification. Their analy
multiple indicators did not reduce the error in c
and science. We suspect that the increasing bur
sions makes it more difficult to accurately diff
attempted to determine who was an engineer b
education, and job history. The results of a mu
indicated that at least 15% of those who were
be missed using various classification schemes

Latour (1987) used the term "technoscienc
between engineering and science. Using a net
scribed the daily activities of both scientists an
sonal success in technoscience did not dep
engineers and scientists performed their jobs, b
to recruit others into believing in the value o
technoscience, recruiting others included writi
ing for projects, doing research, and other activ
ered either science or engineering. That is, succ
does not depend so much on what is made (eng
of new knowledge (scientists) but rather on how
tists are able to recruit others into the process c

When one examines engineers and scientists
it becomes increasingly difficult to distinguish t
tivities that we traditionally consider the activi
(making new products and new knowledge, resp
behave quite differently. Yet many of their acti
the same. Contradictions based on the various vi
the groups contribute to the misunderstanding th
entists.

Differences

Despite the changes in engineering and scien
differences noted by Ritti (1971) still distinguish
engineers in industry, Ritti found marked cont
neers and scientists–(a) the goals of engineers in
with meeting schedules, developing products th

ketplace, and helping the company expand its activities; (b) although both engineers and scientists desire career advancement or development, advancement for the engineer is tied to activities within the organization, whereas advancement for the scientist is dependent upon the reputation established outside the organization; and (c) whereas publication of results and professional autonomy are clearly valued goals of the Ph.D. scientist, they are clearly the least valued goals of the baccalaureate engineer (Ritti cited in Allen 1977, p. 5).

Blade (1963) states that engineers and scientists differ in training, values, and methods of thought. In particular, in their individual creative processes and in their creative products–(a) scientists are concerned with discovering and explaining nature; engineers use and exploit nature; (b) scientists search for theories and principles; engineers seek to develop and make things; (c) scientists seek a result for its own end; engineers are engaged in solving a problem for the practical operating results; and (d) scientists create new unities of thought; engineers invent things and solve problems. Danielson (1960) found that engineers and scientists are fundamentally different in terms of how they approach their jobs, the type and amount of supervision they require, the type of recognition they desire, and their personality traits.

Allen (1977) stated that the type of person who is attracted to a career in engineering is fundamentally different from the type of person who pursues a career as a scientist. He wrote that:

> Perhaps the single most important difference between the two is the level of education. Engineers are generally educated to the baccalaureate level; some have a master's degree, while some have no college degree. The research scientist is usually assumed to have a doctorate. The long, complex process of academic socialization involved in obtaining the Ph.D. is bound to result in persons who differ considerably in their life views. (p. 5)

According to Allen (1984), these differences in values and attitudes toward work will almost certainly be reflected in the behavior of the individual, especially in the use and production of information.

Much of the research on the differences between engineers and scientists is dated and does not reflect the impact of changes in post-World War II engineering curricula. During World War II and throughout the era of Sputnik, government and industry leaders recognized that engineering training in the U.S. was not adequate to meet military and industrial challenges (Grayson, 1993). The Grinter Report, prepared by a committee of the American Society for Engineering Education (ASEE), urged the inclusion of more science and liberal arts into engineering education. This 1955 report transformed engineering education

over the subsequent two decades from "hands-o
perspective resembling other types of academic
ences. In his history of engineering education in
the period from World War II through 1970 the '
education since the 1960s has tended to blur the
of engineers and scientists. In addition, the types
bureaucratic organizations that employ them m
differentiate them by title alone. From a research
serve a clear difference between engineers and s
in this article, we demonstrate the differences b
that are clearly reflected in their daily activities.

Engineering is defined as the creation or in
such, it clearly encompasses both intellectual
knowing and doing). Engineering work is fun
technical activity. It is a social activity in that it
dividuals are required to coordinate and integra
activity in that the production of the final pro
maintain successful social relationships (e.g., ne
smooth personal relations among members of a
community is important for the effective func
and engineers. Engineers do their work in an er
tionships. Science, on the other hand, allows s
activities with only a vague reference to others

Similarities

A number of writers note that engineers beh
At times, they adopt the methods used by scienti
example, according to Ritti (1971), engineering
perimentation, mathematical analysis, design a
ing of prototypes, technical writing, marketi
Kemper (1990), too, noted that the typical eng
lems, come up with new ideas, produce design
work of others, produce reports, perform calc
ments. Florman (1987) described engineering w
ory and empiricism. Ziman (1984) wrote that:

> Technological development itself has bec
> satisfactory, in the design of a new auto
> thumb, cut and fit, or simple trial and erro
> ena are observed, hypotheses are propose
> true spirit of the hypothetico-deductive m

Constant (1980) also described the similarities between engineering and science in his detailed history of the origin of the modern jet engine. He defined a "variation-retention" model to describe how engineers and scientists create technological change. Change, in technology, results from random variation and selective retention. Technological conjecture, which can occur as a result of knowledge gained from either scientific theory or engineering practice, yields potential variations to existing technologies. For example, in the case of the turbojet revolution, technological conjecture was based on engineers' knowledge of scientific theories. In contrast, in their writings, scientists usually describe their methods as following the hypothetico-deductive method. However, in many of their daily research activities, they use methods similar to those used by engineers–particularly the variation-retention method.

Convergence

We expect that many of the differences found by earlier researchers between engineers and scientists should decrease over time. The previous research on the differences and similarities between engineers and scientists has been vague and often based on small samples. Many studies that focused on the differences included engineers who received training before the impact of the Grinter report. In addition, undergraduate engineering curricula are continually changing and are very similar to undergraduate curricula for scientists in that engineers receive more humanities, liberal arts, and business training. As a result, there may be a convergence of the training of each group.

During this century, and especially since World War II, engineers and scientists have been increasingly employed in such large bureaucratic organizations (Florman, 1987; Layton, 1974; Meiksins and Smith, 1993) as the major corporations and the federal government. The integration of engineers and scientists into these organizations has significantly reduced their autonomy. Both groups have increased attempts to maintain their autonomy by defining and controlling separate spheres of knowledge. Yet, in most organizations, the opportunity for upward mobility is limited to management. Both engineers and scientists tend to move into management during their careers. If we look at both groups over their careers, we see that they tend to converge in their daily activities. Although they may consider themselves engineers or scientists based on education or professional orientation, in reality they become managers and behave alike in bureaucratic organizations.

Scientific and Technological Communities

There are other differences between engineers and scientists, in addition to their daily activities, that affect our understanding of their production and use of knowledge. Both engineers and scientists wo that influence their behavior. An analysis of th critical to understanding the production, transfe edge. Engineers and scientists interact within e tematic ways that produce differences in the me access knowledge.

Each group conducts its work activities in ments. For the most part, scientists work within marily "outside oriented." Although scienti proprietary information as part of their duties, t become known to others outside the organizat ment, most scientific work is aimed at others o neers' work is usually more "inside oriented"; th of their organizations. Yet engineers, too, interac their organizations (Kennedy, Pinelli, and Barc

But, as in other ways, the differences between distinct. It is generally accepted that scientists h they use to share information. This "college" ext their work organizations. Within the "college," in among its members. It is assumed that engineers engineers have might better be described as havi with which they share information and ideas (Vin group of people who maintain social contact with mon goals, behavioral norms, and knowledge. A gineers share common knowledge and a set of es prescribes its own approach to work behavior. En ity; most work is accomplished as a result of grou interpersonal communication both within and ou

Studies of scientific communities look at th methods, reward system, and culture shared by quently underscore the role of interpersonal co community and holding it together (see Barb McClure, 1991; Kuhn, 1970). This type of inv performed in relation to engineering knowledge noted that ["the problem of the internal workin nity"] is virtually unexplored. . . . In contrast to community, "little is known about the sociolog neering] community" (p. 495). Constant (1984) on engineering communities. He wrote that "whi done on invisible colleges, research fronts, and t

ence, there has been little analogous, sociological or historical investigation of [engineering] practice" (p. 8).

Rothstein (1969), pointing to the diversity inherent in engineering, warned that defining the entire profession of engineering as a single knowledge community provides a model that is inadequate to describe engineering behavior. He argued that the huge variety of occupations and disciplines in engineering demonstrates that there is no such thing as a single engineering knowledge community. Further, he contended that most discussions of professional communities fail to direct enough attention to the nature of professional knowledge and its influence on behavior. He further contended that the heterogeneity, rate of change, and degree of specialization of engineering knowledge also led to the emergence of specific communities in engineering.

Some work has begun to explore the extent to which members of engineering knowledge communities share similar work tasks, goals, and methods; are governed by shared social and technical norms; and engage in extensive informal information exchange among themselves. Laudan (1984) found justification for this approach in that:

> Cognitive change in technology is the result of the purposeful problem-solving activities of members of relatively small communities of practitioners, just as cognitive change in science is the product of the problem-solving activities of the members of scientific communities. (p. 3)

Layton (1974) also contended that "the ideas of technologists cannot be understood in isolation; they must be seen in the context of a community of technologists" (p. 41). Donovan (1986) noted that "the study of engineering knowledge must not be divorced from the social context of engineering" and suggests that "the interplay of social values and theoretical understanding in the evolution of scientific disciplines certainly has its analogues in engineering, although the values and knowledge involved are often quite different" (p. 678).

The nature of engineering work suggests that engineers require access to a variety of tools and information resources. In addition, the use of these tools and resources and the way they are integrated into engineering work may be planned in some cases and ad hoc in other situations. The engineering knowledge community, although it has received minimal attention from researchers, clearly plays an important role in the conduct of engineering work and the production, transfer, and use of engineering knowledge.

Yet the notion of community can be pushed too far and can cause problems for understanding how engineers behave in their daily activities. As Allen (1977) points out, there is a vague limit on the amount of interpersonal communication engineers can use in problem-solving. It is acceptable for engineers to

ask for help from other engineers when they hav
also expected that engineers will not seek help f
problem they encounter. The normative system
from asking for all the help they might need bec
lack of engineering competence. When the engin
terpersonal communications can no longer be use
edge, testing, or finding the information–become

Differences Between Engineers and Scientists

Most research on knowledge production, trans
the differences between engineers and scientists. R
on how scientists create and use knowledge, or it h
similar. Engineers and scientists are similar in that
is critical to the performance of their jobs, but ther
and when they use it. To thoroughly understand th
diffusion, research needs to be focused more clos
and scientists differ in their production and use of

Scientists use knowledge as part of the proc
edge. Latour (1987) described how scientists "re
search predecessors to demonstrate the importa
They use knowledge to show how their research
previous research in the field. The intellectual c
established if the priority and importance of a fi
lished. In most instances, scientists gather most
ginning research or at least before writing their

Engineers, on the other hand, use knowledge
care more about the ability of the research to prov
lar problem than about its intellectual history. Th
the research, design, development, and manufac
neers produce new knowledge, they often do so
others in their organization who might face a sin
of another product. In these instances, the intell
search is not as important as documenting the pr

Scientists tend to use hypothesis testing (at l
search) in gaining new knowledge. Engineers ar
parameter variation and selective retention (Vin
knowledge. Each technique produces different t
nal articles are appropriate for scientists to descr
ing of one idea. In contrast, technical report
engineers to document engineering outcomes. W

scientists will search for knowledge that may not be directly related to the research but can be used to place the research in context. The information needs of engineers are more immediate, at least in critical phases of design. They select information because it directly relates to solving a problem. For example, according to Allen (1977):

> Engineers read less than scientists, they use literature and libraries less, and they seldom use information services which are directly oriented to them. They are more likely to use specific forms of literature such as handbooks, standards, specifications, and technical reports. (p. 80)

What an engineer usually wants, according to Cairns and Compton (1970), is "a specific answer, in terms and format, that is intelligible to him–not a collection of documents that he must sift, evaluate, and translate before he can apply them" (pp. 375-376). Young and Harriott (1979) report that:

> The engineer's search for information seems to be based more on a need for specific problem solving than around a search for general opportunity. When engineers use the library, it is more in a personal-search mode, generally not involving the professional (but nontechnical) librarian. (p. 24)

Young and Harriott conclude by saying:

> When engineers need technical information, they usually use the most accessible sources rather than searching for the highest quality sources. These accessible sources are respected colleagues, vendors, a familiar but possibly outdated text, and internal company [technical] reports. He [the engineer] prefers informal information networks to the more formal search of publicly available and cataloged information (p. 24). We are not convinced that there is a neat dichotomy between engineers and scientists in their production, transfer, and use of knowledge. Rather, there is a continuum of activities and behaviors that each group uses in differing amounts in their daily activities. Included among these are knowledge production, transfer, and use activities that appear to be similar to those of the other profession. Because of the variety of tasks that engineers and scientists perform, it is difficult to assume that any model of knowledge diffusion can be simple and meet the needs of all engineers and scientists.

The Nature of Engineering Knowledge

Vincenti (1990) proposed a schema for engineering knowledge that categorizes knowledge as descriptive (factual knowledge), prescriptive (knowledge of the desired end), or tacit (knowledge that canno tures but is embodied in judgment and skills). B edge are embedded in the daily activities and These types of knowledge guide their everyday signing products and services. Descriptive know in commonly shared knowledge. The commun sources of knowledge as its form of collective i

Vincenti (1990) traced five "normal" (as opp ments in the history of aerospace engineering to omy of engineering design knowledge" (p. technological developments require a range of sc knowledge as well as information about social, e mental issues. Vincenti conducted three imp knowledge. One analysis is his own elaboration c of the growth of technological knowledge. Vinc numerous examples from history, that the mech in engineering design include three types of (a) searching past experience to find knowledge ing the identification of variations that have not features thought to have a chance of working; ceived variations to choose those most likely to activities occur in an interactive and disord through physical trials such as everyday use, ex tunnels), simulations, and analytical tests such a proposed designs, calculations, and other means lecting a proposed variation (pp. 247-248).

Descriptive and prescriptive knowledge are e plicit. Both tacit and prescriptive knowledge are ing how" (p. 197-198). Vincenti defined sp categories: fundamental design concepts, criteri tools (i.e., mathematical methods and theories an tative data, practical considerations, and desig dural knowledge and judgmental skills) (pp. 2 sources of engineering knowledge that include t tion by engineers during invention, theoretical an search, design practice, production, or direct tria

Vincenti (1992) and others use a "demarcati knowledge (Downey and Lucena, 1995); that knowledge as different from scientific knowled categories of knowledge that Vincenti attribut but the two professions might differ in the value

scientists may not assign as much importance to tacit knowledge as engineers do. Using another model, the network actor model, to understand the differences between engineers and scientists would place an important consideration on the differences between the groups in their production and use of knowledge. Both groups use substantial amounts of knowledge, but each uses it in different ways to recruit others into accepting the merit of their designs (engineers) or ideas (scientists). This difference implies that the knowledge diffusion system must meet differing needs if it is to be used successfully by both groups.

Engineers as Information Processors

The ultimate goal of engineering is to produce a design, product, or process. It is informative to view engineering as an information-processing system that uses knowledge to reduce work-related uncertainty. That is, engineers are heavy information processors. The concept of engineering as an information-processing activity represents an extension of the arguments developed by Tushman and Nadler (1980) and has its roots in open systems theory developed by Katz and Kahn (1966). Throughout the engineering process, data, information, and knowledge are acquired, produced, transferred, and used. The fact that these data, information, and knowledge may be physically or hardware encoded should not detract from the observation that the process of engineering is fundamentally an information-processing activity.

Uncertainty, defined as the difference between the information possessed and the information required to complete a task, is central to the concept of engineering as an information-processing activity. Rogers (1982) stated that coping with uncertainty is the central concept in information use behavior. The process of engineering is one of grappling with the unknown. These unknowns or uncertainties may be technical, economic, or merely the manifestations of personal and social variables. When faced with uncertainty, engineers typically seek data, information, and knowledge. In other words, data, information, and knowledge are used by engineers to moderate technical uncertainty. Because engineering generally entails coping with a relatively high degree of uncertainty, engineering can certainly be viewed as an informational process. Consequently, information-seeking behavior and patterns of technical communication cannot be ignored when studying engineers.

In Orr's (1970) conceptual framework, the engineer is an information processor. This framework focuses on information-seeking behavior and assumes that an internal, consistent logic governs the information-seeking behavior of engineers, individual differences notwithstanding. A project, task, or problem that precipitates a need for information is central to the conceptual framework. This need for information may be internally or exter by Orr as inputs or outputs, respectively. Orr, ci Menzel (1964), Storer (1966), and Hagstrom (1 within the mind of the individual engineer and in knowledge needed to keep up with advances i one's professional duties; to interact with peers, to obtain stimulation and feedback from them. O external stimulus or impetus and serve a variety ing to a request for information from a superv league; reporting progress; providing advice; r advocating; and proposing. Inputs and outputs r and types of data, information, and knowledge.

The conceptual framework for our research ect, task, or problem successfully, specific kind and knowledge are needed. In response, engine sic alternatives: They can create the informatio servation, or other accepted engineering proc search the existing information sources to dete rently available and usable. If they act rationally the information is influenced by three factors–(the relative likelihood of success in acquiring time allocated for addressing the project, task, c of the relative cost (money and/or effort) of the ticipated acceptance of their resolution of the pr ers, other engineers, contractors).

If a decision is made to search the existin choose between two information channels. Th formation through informal methods such as in tions with peers, coworkers, colleagues, gate "key" personnel, and supervisors, or the use of formation. They might also choose to use th which includes libraries, technical information cal information specialists, information produ tion storage and retrieval systems. It is assume particular information channel is influenced by acteristics. Other factors, such as the previous channel, will further influence the channel sel 1968; Orr, 1970; Rosenberg, 1967).

More recent work highlights the value of e tional factors related to information seeking an cal investigation of information use environme

of understanding the context in which information is sought, conveyed, and applied. Context for professional groups, including engineers, is defined as a combination of the nature of work problems, solutions, and settings associated with particular types of jobs. Context for engineers also includes the engineering community (i.e., the methods that other engineers would use to solve a problem). We might add that engineers also consider their community when making these decisions; that is, they consider how the engineering community would approach the problem and the perceived acceptability of their choice of action in the community. Taylor assumes, in other words, that members of a profession share tasks, goals, and needs in a way that influences their use of information. Taylor's analysis recognizes that information-seeking behavior and use are determined by the nature of the particular project, task, or problem at hand (pp. 217-255).

A shift in emphasis toward the study of cognitive and situational factors surrounding information seeking and use, and away from users' personal characteristics and specific systems features, has been advocated by a number of communications and information science researchers, most notably Dervin and Nilan (1986). They have devoted special attention to understanding what there is about a particular situation that encourages an individual to use networks (i.e., group of individuals) in fulfilling an information need. The subjective perception of cost, time, and likelihood of success may often be situationally driven.

The data, information, and knowledge that result from an engineer's search are evaluated subjectively. The engineer as an information processor faces three possible courses of action: first, if the created or available data, information, and knowledge used to complete the project or task or solve the problem are sufficient, the process is terminated; second, if the created or available data, information, and knowledge are useful but only partially sufficient to complete the project or task or to solve the problem, a decision is made either to continue the process by reevaluating the information source selected or to terminate the process; and third, if the created or available data, information, and knowledge are not applicable to or do not complete the project or task or solve the problem, a decision is made either to continue the process by redefining the project, task, or problem or to terminate the process. Throughout the process, the engineer evaluates both process and outcomes in light of what others in the engineering community would do and also in light of the anticipated acceptance by others within the engineering community and the employing organization. The complexities of the decision processes used by engineers to evaluate knowledge require an understanding of the personal, situational, contextual, and community characteristics in which the engineer works.

Major Empirical Studies of Engineering Infor

Pinelli (1991), Pinelli, Bishop, Barclay, an(Casto, and Jones (1994) reviewed many of the s behavior by engineers and scientists. For the mo and reports produced as part of this project, the p on the production and use of knowledge by scie tinguish between engineers and scientists. Bec cused primarily on the information use behavior at length. Overall, there is a remarkable consis neers use information to solve a problem. They when it is needed. They choose informal source believe these sources are effective and efficient

Herner

Herner's (1954) work was one of the first " cerned with "differences" in information-seeki nificant differences in terms of researchers pe research, researchers performing "academic a their information-seeking behavior. Herner stat "basic research" or "academic" duties made gre channels or sources, depended mainly on the lit rial, and maintained a significant number of co tion. Researchers performing "applied researc greater use of informal channels or sources, dep tions of information and colleagues for inform use of the library than did their counterparts, a outside of the organization. Applied or industry use of handbooks, standards, and technical repo less of their reading in the library than did their c

Rosenbloom and Wolek

In 1970, Rosenbloom and Wolek published large-scale industry studies that was specifically of knowledge within R&D organizations. They fundamental differences between engineers an to make substantially greater use of informatio tion than do scientists; (b) scientists make consi fessional (formal) literature than do engineers

likely than are engineers to acquire information as a consequence of activities directed toward general competence rather than at a specific task.

Considering interpersonal communication, the engineers in the Rosenbloom and Wolek study recorded a higher incidence of interpersonal communication with people in other parts of their own corporation, whereas scientists recorded a greater incidence of interpersonal communication with individuals employed outside their own corporation. When using the literature, engineers tended to consult in-house technical reports or trade publications, while scientists made greater use of the professional (formal) literature.

Rosenbloom and Wolek also reported certain similarities between engineers and scientists. The propensity to use alternative types of technical information sources is related to the purposes that give meaning to the use of that information. Work that had a professional focus drew heavily on sources of information external to the user's organization. Work that had an operational focus seldom drew on external sources, and relied heavily on information that was available within the employing organization. Those engineers and scientists engaged in professional work commonly emphasized the simplicity, precision, and analytical or empirical rigor of the information source. Conversely, those engineers and scientists engaged in operational work typically emphasized the value of communication with others who understood and were experienced in the same real context of work.

Allen

Allen's (1977) study of technology transfer and the dissemination of technological information within the R&D organization was the result of a 10-year investigation. Allen described the study, which began as a "user study," as a systems-level approach to the problem of communication in technology. Many information professionals consider his work to be the seminal research on the flow of technical information within R&D organizations. Allen was among the first to produce evidence supporting different information-seeking behavior for engineers and scientists. These differences, Allen noted, led to different philosophies and habits regarding the use of the technical literature and other sources of information by engineers. The most significant of his findings were the relative lack of importance of the technical literature in terms of generating new ideas and in problem definition, the importance of personal contacts and discussions among engineers, the existence of technological "gatekeepers," and the importance of the technical report. Allen stated that "the unpublished report is the single most important informal literature source; it is the principal written vehicle for transferring information in technology" (p. 91).

Kremer

Kremer's (1980) study was undertaken to gain
formation flows through formal and informal char
sign company. The need to solve a problem
frequently to search for information. Kremer fou
company were contacted first for needed informat
the company were contacted. In terms of the te
were most important, followed by standards and
not important sources of information and were u
engineers. Regardless of age and work experien
strated a decided preference for internal sources o
personal files for needed information. The percei
technical quality, and amount of experience a desi
mation source strongly influenced the selection of
nological gatekeepers appeared among design en
high technical performers, and a high percentage

Shuchman

Shuchman's (1981) study was a broad-based
transfer in engineering. The respondents represe
lowing major engineering disciplines: civil, elec
chemical and environmental, and aeronautical. Th
cipline, displayed a strong preference for informa
ther, these engineers rarely found all the inform
technical problems in one source. The major diffic
finding the information they needed was identifyi
data and then learning who had it. In terms of inf
technical problems, Shuchman reported that engi
sonal stores of technical information, followed in
with colleagues, discussions with supervisors, us
and contact with a "key" person in the organizatio
needed information was located. A small proport
sion used technical libraries and librarians. In gen
gineers do not regard information technology a
process of knowledge production, transfer, and
gatekeepers appear to exist across the broad ran
their function and significance are not uniform. C
neering, gatekeepers accounted for only a small p
process.

Kaufman

Kaufman's (1983) study was concerned with the factors relating to the use of technical information by engineers in solving problems. The study reported that, in terms of information sources, engineers consulted their personal collections first, followed by colleagues, and then by formal literature sources. In terms of the formal literature sources used for technical problem solving, engineers used technical reports, followed in order by text books and technical handbooks. Most sources of information, according to Kaufman (1983), were found primarily through an intentional search of written information, followed by relying on personal knowledge and then by asking someone. The criteria used in selecting all information sources, in descending order of frequency, were accessibility, familiarity or experience, technical quality, relevance, comprehensiveness, ease of use, and expense. Engineers used various information sources for specific purposes. They primarily utilized librarians and information specialists to find leads to information sources. Engineers used online computer searches primarily to define the problem, and technical literature to learn techniques applicable to dealing with the problem. They relied primarily on personal experience to find solutions to the problem.

Kaufman (1983) reported that the criteria used in selecting the most useful information sources, in descending order of frequency, were technical quality or reliability, relevance, accessibility, familiarity or experience, comprehensiveness, ease of use, and expense. In terms of the effectiveness, efficiency, and usefulness of the various information sources, personal experience was rated the most effective source of information. Librarians and information specialists received the lowest rating for efficiency and effectiveness. Most engineers used several different types of information sources in problem solving; however, engineers depended on their personal experience more often than on any other single specific information source.

Differences Among Types of Engineers

Just as a careful analysis of engineers and scientists requires that we examine them as separate groups when analyzing some issues, it is equally important to recognize the differences among types of engineers when we look at engineers' attitudes and behaviors related to knowledge production, transfer, and use. Researchers rarely recognize that engineers perform such varied duties as engineering research, design, and production engineering. The daily activities of some engineers are not readily distinguishable from those of scientists. For other engineers, there are few similarities in their daily activities to those of scientists. A critical factor in the analysis of engineering information needs is the possibility of differences amo varied duties that engineers perform.

It would be almost impossible to answer the engineer is and what an engineer does. For the and scientists are based on self-identification c scriptions of the myriad of engineering behavi earlier research indicate that the activities perfo and multifaceted. However, we found no studies ysis of the daily activities of engineers.

The diversity of engineers implies a diversity search not only assumes that scientists and eng tion needs but also attempts to determine whe scientists differ and in what ways. For example, search engineers differ from those of productio scientists are heavy information users, but they the same mix of information. A knowledge di differences and meets the needs of each.

The more "science-oriented" model of en need for additional training in using informat neers' skills training. Two conflicting forces, t neering and the bureaucratization of engineeri engineering during this century. One, engineeri fessional occupation–similar to medicine and l gineers were academically trained, especial land-grant universities. Baccalaureate degrees method of entering the profession, although K 15% of the engineers in her study did not have training, too, changed from a "hands-on" app professional training. At the same time, the ma chemical firms and other large bureaucratic o larger numbers of engineers. Two, these organi the professional activities and independence of technical rather than professional employees Whalley, 1986; Zussman, 1985).

The desire by the leaders of engineering so neering and engineers' feelings that they were workers led to a "revolt of the engineers" duri (Layton, 1974). Engineers became more orient to their employers. Increasingly, engineers f rather than an organization. That is, engi "stand-alone" profession whose skills could b ting to another. The independence of engineers increased their reliance on shared information as critical to their professional identity. Increased reliance on knowledge, along with the changes in academic curricula, contributed to an increased recognition by engineers that the ability to use knowledge was essential to their professional careers, perhaps more so than hands-on training, because it made them less dependent on the training provided by their employers. The sharing of engineering knowledge, as it is contained in information, is the key link among engineers as they define themselves as professionals.

At the same time, the increased concentration of engineers in large bureaucratic organizations, along with the increasing amount of information in engineering knowledge, created the need to develop specialties within engineering that went beyond those tied to academic curricula. In U.S. industries, and in particular the U.S. aerospace industry, multiple types of engineers perform a variety of engineering tasks that range from basic science through engineering science to design and development and to manufacturing and production engineering. We expected to find differences among these subdisciplines in the types of information engineers used, the sources used to find information, and the amount of information they needed to perform their duties. These differences are important to understanding of effective diffusion of aerospace knowledge.

The Aerospace Engineering Community

Our research is unique among engineering studies in that we focus on a single engineering discipline and its many subdisciplines, and on one industry. Previous research on engineers and scientists has generally focused on one organization, or it was widespread across engineering and science disciplines. The focus on an organization can confuse the organizational characteristics with the characteristics of the engineers and scientists. The large multidisciplinary engineering studies do not provide sufficient focus for understanding one engineering discipline thoroughly (e.g., aerospace). We contend that a more effective analysis of engineers must consider the various technological tasks performed by aerospace engineers and scientists, which we describe as a continuum from applied (engineering) science through engineering research to design and development, production engineering, and marketing and sales. At each step, the tasks become more directly related to the production of an aerospace product. At each step along this continuum, the production, transfer, and use of knowledge may differ. This lack of attention to differences has probably contributed to the lack of an effective aerospace knowledge diffusion system.

Aerospace employs a wide range of engineers and scientists who represent many engineering specialties and scientific disciplines. Most of the scientists in aerospace are employed in the national labs and in universities, but some are

employed in major aerospace firms. In contrast,
vate companies that range from major firms lik
firms with a few employees. Aerospace is furthe
gineers in one discipline–aerospace engineeri
reaucracies. As such, aerospace engineers hav
that affect their production, transfer, and use of

The notion of an aerospace engineering c
Vincenti (1990). He described informal comm
most important source of knowledge generat
transfer in aerospace. Vincenti defined a comm
on a particular aerospace development or prob
propellers), and he attributed several functions
ties. Competition among members supplies r
provides mutual support. The exchange of kno
ates further knowledge, which is disseminated b
and teaching, and is also incorporated into the
munity also plays a significant role in providin

Vincenti (1990) described the particular rol
space engineering institutions, such as gover
university departments, aircraft manufacturers,
fessional societies, government regulatory ager
ponent suppliers (pp. 238-240). He concluded,

> Formal institutions do a complex multitu
> channel the generation of engineering k
> ever, constitute the locus for that generati
> mal communities do. Their role . . . is to su
> such communities. (p. 240)

Constant (1980) described aerospace enginee
locus of technological cognition. He noted that th
fact, composed of a multilevel, over-lapping hi
argued that technological change is better studie
the individual, organizational, national, or indust
community as the embodiment of traditions of pr
of practice define an accepted mode of operation
complishing a specified technical task. Such trad
evant scientific theory, engineering design form
methods, specialized instrumentation, and, ofter
nale. A tradition of technological practice is p
community that embodies it; each serves to defi

tice are passed on in the preparation of aspirants to community membership. A technological tradition of practice has, at minimum, a knowledge dimension that includes both software and hardware, and a sociological dimension that includes both social structure and behavioral norms.

In *The Origins of the Turbojet Revolution*, Constant (1980) discussed further the importance of community norms in engineering. He alleged that, at least in connection with complex systems, there are "fundamental social norms governing the behavior of technological practitioners which are very close in structure, spirit, and effect to the norms governing the behavior of scientists" (p. 21). Such norms guide the development of techniques and instruments and the reporting of data. Constant also argued for the existence of "counternorms" in engineering that are similar to those attributed to scientists by Meiksins and Smith (1993). Constant explains:

> Technological practitioners are required to be objective, emotionally neutral, rational, and honest. Yet technological practitioners often are–and protagonists of technological revolution usually are–passionate, determined, and irrationally recalcitrant in the face of unpleasant counter evidence bearing on their pet ideas. (p. 24)

Aerospace engineering might be thought of as a series of communities. Despite being a relatively new engineering discipline, aerospace engineering has diversified as it has grown. The aerospace engineering communities include a range of activities from basic science through very applied production engineering. The communities are held together because of a common use of aerospace-related knowledge. Data demonstrate that aerospace engineers and scientists have varied duties and responsibilities and, consequently, differing information-seeking behaviors and information needs. These various behaviors and needs must be taken into account in the development of an effective system for diffusing aerospace knowledge.

CONCLUSIONS

We described the relationship between science and technology and the similarities and differences between engineers and scientists, and we demonstrated that there is a tendency for convergence in their duties and responsibilities. Self-identification as engineers and scientists remains a definitional problem, and the outputs from both groups are often quite similar. For example, engineers and scientists work together and do similar work in the national laboratories and large industrial research institutions. In aerospace in particular, there are likely to be fewer distinctions between scientists and en dustry is focused on the development and exploit prove flight.

We find ourselves somewhat in agreement and technology are not fully divergent. We arg clear distinction further supports our contentio edge community exists. We believe that the fact that engineers have and use is not clearly disting used in science (Layton, 1992). Rather, the eng nity is based on a shared understanding among e and theoretical information. That is, engineerin neers' understanding how to build into a design onstrate engineering. The engineering knowledg understanding of the engineering built into a de

To give an example of the differences betwee could assume that aerospace includes the follow ics, structures, and propulsion. Scientists work their work on the factual and theoretical foun components. In contrast, engineers would exam nents but would also include one or more of the ogy, psychology, market analysis, systems an management issues–in their work. Because of t dustry, many aerospace engineers and scientists as engineering in their everyday activities. We part of engineers' work (the social part) distingui the inclusion of the social parts of their work t The ability of engineers to understand the socia guishing feature of the engineering knowledge

At the junction of (a) science and technology engineering knowledge community, and (c) an production, transfer, and use, we can best unders aerospace. Science and technology studies de come integrated into the existing science or eng few rare cases, transform the knowledge comm the engineering knowledge community demonst cal innovation) happens in the daily workplac vances in aerospace knowledge cannot happen production, transfer, and use.

Our analysis of existing research into the us coupled with research results from the NASA Diffusion Research Project, allow us to use aero as a means of developing similarities and differences between engineers and scientists in terms of their information-seeking behavior. Collectively, these results confirm that libraries and librarians, as information intermediaries, serve a vital role in completing the producer-to-user transfer of knowledge and in providing vital information to users. Considering that libraries are service organizations, it might be instructive for librarians to examine existing policies and practices as they pertain to the provision of information services to engineers. Do these policies and practices amount to a "one size fits all mentality" in terms of providing information services to engineers and scientists? On the other hand, how active or aggressive can or should librarians be in today's world of constantly decreasing and the wholesale cutting of library resources? Nevertheless, a reading of this article should at least give librarians "cause to think" about the quality of services provided to engineers and the strategies being developed and applied as they attempt to meet the information needs of engineers. If a reader takes just one point from this article it should be that "knowing your customer is fundamental and essential to servicing the information needs of your customer."

REFERENCES

Allen, T.J. (1984). "Distinguishing Engineers From Scientists." In *Managing Professionals in Innovative Organizations*. R. Katz, ed. Cambridge, MA: Ballinger Publishing, 3-18.

Allen, T.J. (1977). *Managing the Flow of Technology: Technology Transfer and the Dissemination of Technological Information Within the R&D Organization*. Cambridge, MA: MIT Press.

Amabile, T.M. (1983). *The Social Psychology of Creativity*. New York, NY: Springer-Verlag.

Amabile, T.M., and S.S. Gryskiewicz. (1987). *Creativity in the R&D Laboratory*. Greensboro, NC: Center for Creative Leadership.

Barber, B. (1962). *Science and the Social Order*. rev ed. New York, NY: Collier Books.

Bikson, T.K.; B.E. Quint; and L.L. Johnson. (1984). *Scientific and Technical Information Transfer: Issues and Options*. Washington, DC: National Science Foundation. (Available NTIS; PB-85-150357; also available as Rand Note 2131.).

Blade, M.F. (1963). "Creativity in Engineering." In *Essays on Creativity in the Sciences*. M.A. Coler, ed. New York, NY: New York University Press, 110-122.

Bush, V. (1945). *Science: The Endless Frontier*. Washington, DC: Government Printing Office.

Cairns, R.W. and B.E. Compton. (1970). "The *SATCOM Report and the Engineer's Information Problem*." Engineering Education 60: 375-376.

Citro, C.F. and G. Kalton. (1989). *Surveying the Nation's Scientists and Engineers: A Data System for the 1990s*. Washington, DC: National Academy Press.

Constant, E.W. II. (1984). "Communities and Hierarc Science and Technology." In *The Nature of Techn of Scientific Change Relevant*? R. Laudan, ed. Bo

Constant, E.W. II. (1980). *The Origins of the Turbojet* Johns Hopkins University Press.

Danielson, L.E. (1960). *Characteristics of Engineer Their Motivation and Utilization*. Ann Arbor, MI:

Dasgupta, P. (1987). "The Economic Theory of Techn *nomic Policy and Technological Performance*. P. Cambridge, UK: Cambridge University Press, 7-2

Dervin, B. and M. Nilan. (1986). "Information Need *nual Review of Information Science and Technol* York, NY: John Wiley, 3-33.

Donovan, A. (1986). "Thinking About Engineerin 674-677.

Doty, P.; A.P. Bishop; and C.R. McClure. (1991). " Electronic Research Networks." Paper presented a American Society for Information Science. 28: 24

Downey, G.L. and J.C. Lucena. (1995). "Engineering *of Science and Technology Studies*. S. Jasanoff; G Pinch, eds. Newbury Park, CA: Sage Publications

Florman, S.C. (1987). *The Civilized Engineer*. New Y

Gaston, J. (1980). "Sociology of Science and Techno *of Science, Technology, and Medicine*. P.T. Durbin 465-526.

Gerstberger, P.G. and T.J. Allen. (1968). "Criteria Use Engineers in the Selection of an Information Sourc *ogy* (August) 52(4): 272-279.

Grayson, L.P. (1993). *The Making of an Engineer: neering Education in the United States and Canac*

Hagstrom, W.O. (1965). *The Scientific Community*. N

Herner, S. (1954). "Information Gathering Habits of V ence." *Industrial Engineering and Chemistry* 46(1

Illinois Institute of Technology. (1968). *Technology in in Science*. Washington, DC: National Science Fou (Available NTIS; PB-234767.).

Katz, D. and R.L. Kahn. (1966). *The Social Psycholo* NY: John Wiley.

Kaufman, H.G. (1983). *Factors Related to the Use of neering Problem Solving*. Brooklyn, NY: Polytech

Kemper, J.D. (1990). *Engineers and Their Profession* Saunders.

Kennedy, J.M.; T.E. Pinelli; and R.O. Barclay. (1995). tion-Seeking Behaviors of Three Groups of U.S. A sented at the 33rd Aerospace Sciences Meeting & I of Aeronautics and Astronautics (AIAA), Reno, NV. AIAA 95-0706 (Available NTIS; 95N19127.).

King, D.W.; J. Casto; and H. Jones. (1994). *Communications by Engineers: A Literature Review of Engineers' Information Needs, Seeking Processes, and Use*. Washington, DC: Council on Library Resources.

Kintner, H.J. (1993). *Counting Engineers–A Latent Class Analysis of Self-Reported Occupation, Employer Administrative Records, and Educational Background*. GMR-8033. Warren, MI. General Motors Corporation, NAO Research and Development Center.

Kremer, J.M. (1980). *Information Flow Among Engineers in a Design Company*. Ph.D. Diss., University of Illinois at Urbana-Champaign. UMI 80-17965.

Kuhn, T. (1970). *The Structure of Scientific Revolutions*. 2nd ed. Chicago, IL: University of Chicago Press.

Landau, R. and N. Rosenberg. (eds.). (1986). *The Positive Sum Strategy: Harnessing Technology for Economic Growth*. Washington, DC: National Academy Press.

Langrish, J; M. Gibbons; W.G. Evans; and F.R. Jevons. (1972). *Wealth From Knowledge: A Study of Innovation in Industry*. New York, NY: John Wiley.

Latour, B. (1987). *Science in Action: How to Follow Scientists and Engineers Through Society*. Cambridge, MA: Harvard University Press.

Laudan, R. (ed.). (1984). "Introduction." In *The Nature of Technological Knowledge: Are Models of Scientific Change Relevant*? R. Laudan, ed. Boston, MA: Reidel, 1-26.

Law, J. (1987). "The Structure of Sociotechnical Engineering: A Review of the New Sociology of Technology." *Sociological Review* 35: 405-424.

Law, J. and M. Callon. (1988). "Engineering and Sociology in a Military Aircraft Project: A Network of Analysis of Technological Change." *Social Problems* 35: 115-142.

Layton, E.T. (1992). "Escape From the Jail of Shape; Dimensionality and Engineering Science." In *Technological Development and Science in the Industrial Age*. P. Kroes and M. Bakker, eds. Dordrecht, The Netherlands: Kluwer Academic Publishers, 69-98.

Layton, E.T. (1974). "Technology as Knowledge." *Technology & Culture* 15(1): 31-33.

Meiksins, P.F. and C. Smith. (1993). "Organizing Engineering Work: A Comparative Analysis." *Work and Occupations* 20(2): 123-146.

Menzel, H. (1964). "The Information Needs of Current Scientific Research." *Library Quarterly* (January) 34(1): 4-19.

Orr, R.H. (1970). "The Scientist as an Information Processor: A Conceptual Model Illustrated With Data on Variables Related to Library Utilization." In *Communication Among Scientists and Engineers*. C.E. Nelson and D.K. Pollock, eds. Lexington, MA: D.C. Heath, 143-189.

Pinelli, T.E. (1991). "The Information-Seeking Habits and Practices of Engineers." *Science & Technology Libraries* 11(3): 5-25.

Pinelli, T.E.; A.P. Bishop; R.O. Barclay; and J.M Kennedy. (1993). "The Information-Seeking Behavior of Engineers." In *Encyclopedia of Library and Information*

Science. A. Kent and C.M. Hall, eds., 52:15
167-201.

Price, de Solla D.J. (1965). "Is Technology Histo
Technology & Culture (Summer) 6(3): 553-578.

Rip, A. (1992). "Science and Technology As Dancin
velopment and Science in the Industrial Age.
Dordrecht, The Netherlands: Kluwer Academic

Ritti, R.R. (1971). *The Engineer in the Industrial C*
lumbia University Press.

Rogers, E.M. (1982). "Information Exchange and Te
5 in *The Transfer and Utilization of Technical Kn*
MA: D.C. Heath, 105-123.

Rosenberg, V. (1967). "Factors Affecting the Prefe
Information Gathering Methods." *Information*
119-127.

Rosenbloom, R.S. and F.W. Wolek. (1970). *Techno*
Survey of Practice in Industrial Organizations.
School Press.

Rothstein, W.G. (1969). "Engineers and the Functi
The Engineers and the Social System. R. Pericc
NY: John Wiley, 73-97.

Salomon, J-J. (1984). "What is Technology? The Is
History and Technology 1(2): 113-156.

Shapley, D. and R. Roy. (1985). *Lost at the Fronti*
Policy Adrift. Philadelphia, PA: ISI Press.

Shuchman, H.L. (1981). *Information Transfer in E*
Futures Group.

Storer, N.W. (1966). *The Social System of Science*. N
Winston.

Taylor, R.S. (1991). "Information User Environme
tion Sciences 10. Norwood, NJ: Ablex, 217-255.

Tushman, M.L. and D.A. Nadler. (1980). "Comm
R&D Laboratories: An Information Processing
search and Innovation. B.V. Dean and J.L.
North-Holland, 91-112.

U.S. Department of Defense, Office of the Directo
neering. (1969). *Project Hindsight*. Washington,
(Available NTIS; AD-495, 905.).

Vincenti, W.G. (1990). *What Engineers Know an*
Studies From Aeronautical History. Baltimore,
Press.

Vincenti, W.G. (1992). "Engineering Knowledge, T
archy; Further Thoughts About What Engineers
velopment and Science in the Industrial Age.
Dordrecht, The Netherlands: Kluwer Academic

Voight, M.J. (1960). *Scientists' Approaches to Information.* ACRL Monograph, No. 24. Chicago, IL: American Library Association.

Weingart, P. (1984). "The Structure of Technological Change: Reflections on a Sociological Analysis of Technology." In *The Nature of Technological Knowledge: Are Models of Scientific Change Relevant*? R. Laudan, ed. Boston, MA: Reidel, 115-142.

Whalley, P. (1986). *The Social Production of Engineering Work.* Albany, NY: SUNY Press.

Young, J.F. and L.C. Harriott. (1979). "The Changing Technical Life of Engineers." *Mechanical Engineering* (January) 101(1): 20-24.

Ziman, J. (1984). *An Introduction to Science Studies: The Philosophical and Social Aspects of Science and Technology.* Cambridge, UK: Cambridge University Press.

Zussman, R. (1985). *Mechanics of the Middle Class.* Berkeley, CA: University of California Press.

Current Awareness Reports at Albany International Research Co.

Alex Caracuzzo

SUMMARY. Current awareness services and selective dissemination of information are used at Albany International Research Co. (AIRESCO) to support the research, product development, and other information needs of our scientific and technical community. The nature of our industry, our business objectives, and the variety of information needs pose challenges when compiling timely alerts to newly published information. This paper examines a number of issues affecting the Current Awareness Reports program administered by our one-person science library. *[Article copies available for a fee from The Haworth Document Delivery Service: 1-800-HAWORTH. E-mail address: <docdelivery@haworthpress.com> Website: <http://www.HaworthPress.com> © 2001 by The Haworth Press, Inc. All rights reserved.]*

KEYWORDS. Current awareness, selective dissemination of information, paper machine clothing, engineering, textiles

INTRODUCTION

Current awareness (CA) services and selective dissemination of information (SDI) are used at Albany International Research Co. to support the research,

---

Alex Caracuzzo, MLIS, is Information Specialist, Albany International Research Co., Mansfield, MA.

The author would like to acknowledge the efforts and dedication of former AIRESCO librarians Jeannette Davis and Nora Tillman.

[Haworth co-indexing entry note]: "Current Awareness Reports at Albany International Research Co." Caracuzzo, Alex. Co-published simultaneously in *Science & Technology Libraries* (The Haworth Information Press, an imprint of The Haworth Press, Inc.) Vol. 21, No. 3/4, 2001, pp. 165-173; and: *Information and the Professional Scientist and Engineer* (ed: Virginia Baldwin, and Julie Hallmark) The Haworth Information Press, an imprint of The Haworth Press, Inc., 2001, pp. 165-173. Single or multiple copies of this article are available for a fee from The Haworth Document Delivery Service [1-800-HAWORTH, 9:00 a.m. - 5:00 p.m. (EST). E-mail address: docdelivery@haworthpress.com].

http://www.haworthpress.com/store/product.asp?sku=J122

10.1300/J122v21n03_10

product development, and other information nee cal community. The nature of our industry, our b ety of information needs pose challenges when c published information. This paper examines a nu Reports program administered by our one-perso

HISTORY OF CURRENT AWARENESS

Albany International Research Co. (AIRESC ment facility in Mansfield, Massachusetts and is Albany International Corp. Albany Internationa of engineered textiles called paper machine clo dustry. These large fabrics carry the paper shee Our core business at AIRESCO is to innovate a PMC products for our parent company. We hav ing new business opportunities by applying our to other industries. Thermal protection felts cc orbiter, PrimaLoft™ insulation for the apparel tries, and many other innovative products have

For nearly sixty years, a solo librarian has su of our research scientists and engineers and ma collection of textile, chemistry, and paper indus brarian has historically performed a variety of tr routing journals, delivering tables of contents, nouncements to keep employees current with a est. These traditional CA methods are still emp

In 1998, the library began actively marketin rent Awareness Reports," to help researchers technology trends and spot competitive develo tomers meet with the librarian to create a CA p frequency, and format of the report and to formu may elect to receive updated results as often yearly, depending on their individual informatic can choose to receive the reports in a database, in

Based on this profile, a CA Report is generat basis using proprietary databases and the Internet tations and abstracts from newly published tec ceedings, company news stories, patent filing librarian executes the search strategy, removes du out obviously irrelevant information, formats the results, and delivers them to the recipients. When a customer prefers results to be delivered to an intranet database with e-mail alerts to new information, the librarian makes certain that copyright fees are paid when necessary and advises customers on copyright matters when information is to be shared. Customers frequently share and discuss their results with other researchers on their project teams and in their communities of practice. Although the abstracts in the CA Reports are informative, customers usually request full-text versions of the most interesting abstracts. Requests for full-text documents are submitted to the librarian to be filled by the library's print collection and by a network of document delivery suppliers.

The number of active CA Reports fluctuates as new projects begin and old projects fade. The librarian currently administers eight CA Reports, each requiring one to two hours of the librarian's time to compile, format, and deliver. Thirteen of our company's forty-five "technical" employees (scientists, chemists, engineers, technologists and technicians) subscribe to CA Reports. Another fourteen technical employees in other divisions also subscribe to CA Reports.

OUR CUSTOMERS

Pinelli has suggested that "scientists" prefer to begin researching an unfamiliar topic via secondary sources and are likely to first consult with librarians for help in selecting the appropriate information channels, while "engineers" are more likely to begin their research by consulting colleagues (Pinelli, 1991 and Holmstrom, 2001). It is difficult to divide our CA Report customers into such discrete categories. The cross-functional nature of our researchers' work roles, backgrounds, and individual information-seeking behaviors defies generalizations. At AIRESCO, the titles "engineer," "scientist," "chemist," and "technician" are indistinguishable and irrelevant from an information services standpoint.

The line between technical research staff and management is also quite blurry. Upper level managers are involved in the technical aspects of projects, while junior level employees are also called upon to be project leaders. A recent survey of AIRESCO staff indicated that both are equally likely to first consult the library or colleagues for information in an unknown field.

Senior level managers, who are ultimately responsible for the technical directions of projects and for whom information overload is particularly problematic, also use CA Reports to stay informed. Nearly every senior level manager and many project managers at AIRESCO have received or still receive a CA Report from the library.

CHALLENGES FOR CA REPORTS

The nature of research at our company neces variety of information sources covering many know from where the idea for the next new produ Two issues present a challenge to the libraria technical and business information for our resea source dedicated to tracking the paper machine prevalence of textile- and fiber-related informat

No PMC Database

The PMC industry is dominated by only a ha cluding Albany International. There is no databa PMC information. PMC research has historicall and paper industry literature, and indexed in pr *Textile Technology Digest*, *World Textile Abstra* *(Packaging, Paper, Printing and Publishing,* *stracts).* Textile, paper, materials science, che even biology databases need to be included in n gies to retrieve comprehensive results and impro vant information.

Textile Information Abounds

On the other hand, textile information is as abu textiles themselves are in our everyday lives. Any nology is fair game as a source of ideas, if not as a Albany International. Therefore, we cannot limit bases. The popular press and business journals n periodicals include valuable information about applications of big corporations and of small st technologies that could potentially be applied to o tions– as well as "how they do it"–are of great inte be overlooked if we limit our sources of CA infor industry literature.

THREE USES OF CA REPOR

CA Reports are used for one (or several) of t (1) to generate and "brainstorm" ideas that ma supply information in support of moving a proje opment process; and (3) as a competitive intelligence tool. The reason for requesting a CA Report influences the search strategy as much as the subject matter itself and determines the selection of source databases for the report.

Sparking Innovation

CA Reports are often used as a tool for generating new product ideas. These reports search for technology trends and applications that may be applied to our industry. They may be issued indefinitely and require searching technical databases and Websites in multiple industries. If the customer wants to track applications or properties of a specific material, such as uses for polyesters with a high melting point, interesting innovations are as likely to arise from the consumer apparel industry as from the defense industry literature. Newspapers, press releases, and popular magazines often contain insightful articles and ideas, written neatly and concisely, about new technologies and major players.

Supporting Projects

At the onset of new product development projects, CA Reports are requested from the library in the hopes of finding a "breakthrough" to solve a particular technical problem or to support the business case for proceeding with a project. There is already a vague notion of the information sought. Therefore, the scope of the report tends to be somewhat narrowly defined. The researchers have suggestions for the best places to search for new information, and they are able to identify some of the scientists and companies already involved in the field. This allows us to identify databases with a relatively high probability of finding information germane to the project.

A broad patent technology search helps researchers develop a more complete understanding of relevant prior art, which assists them in the patent filings that must accompany new product projects. Since limiting intellectual property research to our industry's patent classes will not retrieve results in unfamiliar industries, patent searches are usually not limited by patent class codes.

The goal of these projects is, ultimately, commercialization and profitability. Data to support the commercialization plan behind a project is therefore also useful and may require searching business and market research sources as well.

Competitive Intelligence

Secondary sources are not always the best places to find cutting edge information on other companies' activities. Nevertheless, CA Reports are helpful for monitoring some of the technical and strategic activities of other firms. Databases where company names appear freque such as DIALOG's *Company Name Fina* *DIALINDEX* (DIALOG file 411). Company V combined efforts of the librarian and the resea signed to locate patents by targeted company na nies' inventors who may have patent applicatio or to former employers. Press releases and con identify firms and individuals that are active in

A NOTE ABOUT CONTENT

Whatever the primary reason for initiating a quest some combination of technical and com breadth of information required for CA Report tries, DIALOG has been the cornerstone of broad coverage allows us to access databases co technologies of interest to our researchers. Pre librarian by storing and running search strategie such as STN, Lexis-Nexis, and Dow Jones are

Chemical Abstract Society's enduser prod our users' desktops for those who wish to perfo ing. *SciFinder* does not cover general news sou for surveying current patents, but it does allo searches and save successful search queries as ' egies that will automatically run on a regular s

A previous *SciFinder* search is sometimes a initiate a CA Report through the library. Sci formed a *SciFinder* search are better able to exp results they are looking for by demonstrating *SciFinder* users also have a better grasp of the i searching and the complexities of formulating s also tend to have a higher tolerance for multipl

FILTERING AND FORM

In all three uses of CA Reports, many sourc prehensive reports. While some researchers wi essary to digest each and every abstract in a CA few pages of citations before giving up on the

and researcher work together to squeeze as much value from the CA reports as possible by communicating and reaching a mutually agreed-upon conclusion about what level of filtering and volume of results best suits the information-gathering strategy of the researcher. To avoid information overload, the librarian consults with researchers at the onset of the project (and again periodically) to ensure that the search strategy is picking up relevant information. Reports are screened for duplicate information and obvious false hits before delivering to the customer. In many cases, however, false hits are simply unavoidable. The customer's tolerance for false hits and the familiarity of the librarian with the technology at hand determine the level of filtering performed by the librarian before the results are delivered.

In a solo library, time and staff issues are the ultimate factors in determining how much formatting is performed. To maximize the usefulness of CA Reports, and in the interests of value-added customer service, some formatting is always necessary. To enable readers to save time by quickly skimming result sets, highlighting and boldfacing of keywords can be added rather easily using the find-and-replace feature of the text-editing program. Simple headings are added as necessary to divide the report into sections (for example, "company news," "technical papers," "patents," etc.). The librarian occasionally adds brief descriptive notes or abstracts for clarification.

The end result of the filtering process is, hopefully, a CA Report that has the appropriate volume of information and formatting for the intended audience. The librarian must manage the expectations of the CA Report recipients, acknowledge that they aren't going to "catch" everything of interest (de Stricker 2002) and try to resourcefully meet customer expectations despite the limitations of technology and their other responsibilities as a solo librarian.

MARKETING CA REPORTS

CA Reports do not generate tremendous demand by word of mouth, and most of our researchers have never received comparable reports in their prior academic or work experiences. Thus, marketing CA Reports within our company requires a "try it, you'll like it" approach:

- When new employees begin, each is given a library orientation during the first week of employment. This includes a tour of the library and a conversation about the various information services available to employees, including current awareness and SDI services. Employees are shown a sample CA Report and are extended an invitation to meet at greater length to develop a CA strategy around their future projects.

- When a one-time search request is a good Report, the librarian offers to periodically often leads to further discussions with the search strategy.
- Recipients of established CA Reports ar colleagues who might benefit from recei volves more employees in the search fo collaboration and aids the idea generatio
- If the librarian thinks a certain non-custon port, a sample report is generated around th is delivered with an offer to refine and re worthwhile.

Current Awareness Reports are sometimes a try one, they generally stick with the report un Employees then tend to start new reports on oth multiple reports simultaneously.

MEASURING SUCC

CA Reports definitely have a niche in our number of technical employees subscribing to C and in other divisions. We can further conclude from the positive feedback and numerous wor tomers. In many instances, customers have rep to projects being taken in entirely new and succ ers have gratefully reported that CA Reports ha otherwise would not have found without spendi (if they would find it at all). Here are a few out

- A CA Report tracking chemical prices em gotiate a contract with a materials suppli thousands of dollars.
- In competing for a multi-million dollar g search and product development, a CA Rep team members with new patent and market well prepared to present our proposal and v
- A CA Report alerted a project group to a they had not informed us about (it was de We appeared credible, knowledgeable, an velopment efforts when we told them we cently issued.

THE FUTURE OF CA REPORTS AT AIRESCO

We want to repeat the successes of our CA Reports on a larger scale for all R&D researchers and other Albany employees worldwide. Macek notes that current awareness services typically evolve from paper-based, bibliographic, and local services to electronic, full-text, global services (1998). Our CA services have evolved to somewhere past the midpoint on this scale. Integrating full-text and offering other options for global access, perhaps as part of an intranet information portal, is the challenge ahead. Solomon notes that the speed of current awareness information has now been marginalized. Customers expect usefulness, not timeliness (1999). Our focus must remain on customer needs as we continue to develop a useful, flexible and reliable CA Reports program, and as we continue to add value to our company through information services.

REFERENCES

De Stricker, Ulla. "Keep Me Posted . . . But Not Too Much: Challenges and Opportunities for STM Current-Awareness Providers." *Searcher* 10 (January 2002): 52-59.

Holmström, Jonas. "Improving Corporate Library Services." Pro gradu, Åbo Akademi University, 2001: 26-27. <http://www.abo.fi/~jholmstr/publications/improvinglibraryservices.pdf>.

Macek, Rosanne. "Current Awareness at Bay Networks: Some Thoughts on Carbon-Based Filtering." *Special Libraries Association Communications Division Newsletter* (December 1998). <http://www.sla.org/division/dcom/bulletins/sla_1298.html>.

Pinelli, Thomas E. "The Information-Seeking Habits and Practices of Engineers." In *Information Seeking and Communication Behavior of Scientists and Engineers*, edited by Cynthia Steinke, 5-26. New York: The Haworth Press, Inc. 1991.

Solomon, Marc. "When Push Comes to Pull: Serving Current Awareness Applications in Your Company's News Cafeteria." *Searcher* 7 (June 1999): 70-76.

Supporting the Information Needs of Geographic Information Systems (GIS) Users in an Academic Library

Julie Sweetkind-Singer
Meredith Williams

SUMMARY. The growing use of GIS in university research and teaching environments has created a demand for data, software, and technical support that is best accommodated by a central GIS service provider. The library is a natural candidate for this role. While some academic libraries only provide GIS-formatted data, others offer exhaustive services, ranging from GIS courses to contract GIS work for students and faculty upon demand. This paper discusses how the Stanford University Library System has integrated GIS support into its suite of services offered to patrons across all academic disciplines. Two case studies illustrate how the library meets patrons' diverse and detailed GIS information and technical needs. *[Article copies available for a fee from The Haworth Document Delivery Service: 1-800-HAWORTH. E-mail address: <docdelivery@haworthpress.com> Website: <http://www.HaworthPress.com> © 2001 by The Haworth Press, Inc. All rights reserved.]*

Julie Sweetkind-Singer, MBA, MLIS, is GIS and Map Librarian, Branner Earth Sciences Library & Map Collections, Stanford University, Stanford, CA 94305 (E-mail: sweetkind@stanford.edu). Meredith Williams, BA, is GIS Manager, Branner Earth Sciences Library & Map Collections, Stanford University, Stanford, CA 94305 (E-mail: mjwilliams@stanford.edu).

[Haworth co-indexing entry note]: "Supporting the Information Needs of Geographic Information Systems (GIS) Users in an Academic Library." Sweetkind-Singer, Julie, and Meredith Williams. Co-published simultaneously in *Science & Technology Libraries* (The Haworth Information Press, an imprint of The Haworth Press, Inc.) Vol. 21, No. 3/4, 2001, pp. 175-190; and: *Information and the Professional Scientist and Engineer* (ed: Virginia Baldwin, and Julie Hallmark) The Haworth Information Press, an imprint of The Haworth Press, Inc., 2001, pp. 175-190. Single or multiple copies of this article are available for a fee from The Haworth Document Delivery Service [1-800-HAWORTH, 9:00 a.m. - 5:00 p.m. (EST). E-mail address: docdelivery@haworthpress.com].

http://www.haworthpress.com/store/product.asp?sku=J122

10.1300/J122v21n03_11

KEYWORDS. GIS, academic libraries, s

INTRODUCTION

Technological developments are driving char science. One arena of technology that has made a braries' collections and services is Geographic In braries are seeing growing user interest in GIS their staff, services provided, and collection de modate this demand. Each library may choose degree of responding to the users' interests, b manding some reaction from all. Libraries must extensively and to whom they will provide GI other campus groups using and supporting GIS? ware and data does GIS require? How can a lib and faculty to meet their GIS needs? How muc vide support in-house vs. support via the Web aware of the library's GIS support services? T types of questions by describing the GIS support braries (SUL) system.

BACKGROUND

Geographic Information Systems are sets of the storage, retrieval, analysis and display of s Most data contain a geographical component, su street address, city, county, or latitude/longitude of GIS, otherwise disparate data can be related graphic location, creating new information fr Through GIS software, one can display, explore, revealing hidden patterns, relationships, and tren ent in traditional spreadsheet or statistical pack duce high quality static maps with GIS, it is a dy to select and remove any criteria on the map. Thi analysis of the spatial variables affecting a syster ing complicated decision-making.

Early use of GIS was primarily limited to res computer science departments. However, as GI sophisticated, its pool of users has widened. In ur broad range of disciplines are seeing the benefit of this tool for analyzing their spatial data. The advantage of having a unifying GIS platform in which users may combine otherwise disparate data sources is attracting an ever increasing number of users. At Stanford University, students and faculty have used GIS for research in biology, earth sciences, civil engineering, political science, sociology, anthropology, history, electrical engineering and in over 20 other departments and centers across campus. Whether analyzing a digital elevation model, hydrologic network, or census tracts, the benefits of collecting and analyzing information in a GIS are becoming more commonly known.

LIBRARY GIS SERVICES

SUL has adopted the role of being the central node for GIS support on campus. After visiting several other universities' GIS labs, the advantages of housing GIS services in the libraries became clear. At many universities any department with students and faculty interested in GIS is required to provide its own software, data and technical assistance for its GIS users. Often this type of support first develops in a department such as geography, forestry, or earth sciences. Such departments, with large numbers of students interested in GIS, are usually able to satisfy their users' needs, but in departments where only a few students are interested in GIS, often little support is available. Also, this departmental model typically doesn't facilitate sharing of costs and data between the various GIS support nodes that simultaneously develop across the campus. A library's role is to collect and catalogue data such as those needed by GIS users. Providing GIS support through the libraries gives users from all departments equal access to services as the library is often in a central location with open access and long hours of operation.

Within the library system at Stanford the GIS services were initially sited in the Branner Earth Sciences Library and Map Collections (Branner) because it served the departments who first expressed heavy interest in GIS, and it houses nearly all of the cartographic materials for the library system (Derksen et al., 2000). Branner Library supports the School of Earth Sciences, and the earliest users of GIS were from its departments as well as from the Electrical Engineering Department and the Epidemiology Program. Therefore, the GIS support in Branner Library has been tailored to assisting patrons in the sciences. Stanford's Green Library Social Science Resource Center also provides GIS support with a focus on social science and humanities applications. The libraries at Stanford have proven to be a successful central location for providing GIS support to the campus as a whole.

It should be noted that adding GIS services
crease demand on staff resources. This may in
hours, attending advanced training in GIS theory
sistants to maintain Web pages and assist with
skills to manage computers with sophisticated
data. Additional library staff are often required to
A GIS specialist would organize the GIS servic
and provide technical support, while a GIS lib
and become familiar with the large amounts of
produced. The GIS librarian will need to develop
matted data that reduces the duplication of data

GIS SUPPORT AT BR

Patrons come to the library with different lev
in GIS. Some have heard about it but have no cl
ing curve inherent in mastering the software. A
be carried out in order to assess their needs and
to learning a complicated software program. O
map with added points or labels for a report, an i
as Microsoft Encarta or map publishing softwar
will more easily suit their needs.

At Branner a library specialist consults with
cide if GIS is the right tool for their research
each patron's comprehension of GIS and directs
tional materials, and software manuals as nee
abecedarian introduction to GIS, meeting the p
students are very efficient at using and incorpo
als. After the patron has completed one of th
consultation is provided in the form of answeri
general project design or data acquisition assis

Many patrons aren't interested in becoming
like the fastest, simplest way of making the GIS
ticular needs. People exhibiting this learning s
the suggestion of completing a tutorial before as
trons may instead use trial and error methods or
independently to answer their GIS questions (Y
dents, for example, have occasionally found th
instructional materials simplistic and therefore
(Freyberg, 2002). If this introductory GIS stage

are often missed or additional benefits from GIS may not be realized. Also, patrons who skip the self-instruction step often make more demands upon the GIS staff who provide technical assistance. Hence, the staff tries to encourage self-instruction by suggesting the most efficient, high-quality instructional materials available.

Data

Those who come to the library already knowing how to use GIS or willing to learn it may need data, answers to technical questions, and help with project design. As is the case with paper maps, patrons often assume that GIS data in a readily useable format must exist if the data can be conceived. While this may be true for general small-scale data, it is not true with more specific data requests. For example, a student recently requested GIS files showing watershed boundaries, property boundaries, and addresses for a coastal Mendocino county. All of this information can be found, but not in the format desired. If a patron is willing to invest some time in learning GIS, the map layers could probably be created by manipulating Digital Elevation Models, parcel maps, or local government files.

At a large research university the variety of research projects is enormous. Trying to anticipate all of the necessary data becomes an onerous and potentially expensive task. It can be handled more easily by first considering which basic GIS information will serve the largest number of users. It is important to have general data from the Census Bureau for socio-economic data and from the United States Geological Survey (USGS) for elevation data, roads, rivers, and geology. Much of these data are available on the Web for free from government sites or through the Depository Library Program. There are vendors who produce valued-added products that manipulate the data in such a way that it can be more easily handled in a GIS. Commercial sources may also provide the specific data that users request. In selecting products one should consider user demand, budget considerations, and copyright restrictions. Commercial vendors have a wide range of copyright restrictions ranging from fairly lax, where one can use the data on any workstation, to very restrictive, where viewing data is permitted on only one specific machine for the entire campus.

Base map data on a global scale is relatively easy to find, although not always inexpensive. Base map layers include political boundaries, major roads, cities and towns, waterways, etc. Typically, as data resolution improves, costs increase. The Environmental Systems Research Institute (ESRI) Data & Maps CD-ROM set, which is included in an ESRI campus site license, contains base map data for the world, with more detailed data for the United States. The USGS is releasing a collection of CD-ROMs in their Global GIS series that will provide

a world-wide vector data reference at a regional
free to depository libraries that choose to receiv

In comparison, the National Imagery and M
leasing Vector Map Level 1 (VMAP1), which
scale of 1:250,000 in a nonstandard format. A t
ing the data into standard GIS formats. So far, t
tal zones at a price of between $50-95 per zone,
purchased. Some of these costs can be mitigat
ments. The University of California (UC) camp
often purchase expensive data as part of a conso
UC campuses agreed to purchase Landsat 7 im
at $600 per image. The price was leveraged acr
bigger budgets paying for more images. Of c
done if copyright allows it.

In an attempt to influence information-see
Stanford GIS Web site (http://gis.stanford.edu)
data. The first step is to search in Socrates, St
The cataloging records for data received are m
stating that it is "geographic information syste
be used in a keyword search along with a geogr
partial list of GIS datasets held in the library,
another Web page. Next, the patron is encourag
GIS data directories and warehouses, such as th
data clearinghouse sites.

The Internet is often a fruitful place to find d
a keen eye for quality. Search engines, such a
com/), when queried specifically enough, can o
Web sites, like the list archives at Direction
Knowledge Base, allow patrons to search for
post their own query. Patrons are also encoura
staff at Branner who offer searching expertise a
the data available in a variety of places.

Those looking for data do not always emplc
Patrons often prefer to find their data online b
searching on the GIS machines housed at Bran
in order to lower the "cost" of obtaining the infc
lematic if the person only tries to find the right a
needs, say, when looking for a Digital Elevatio
data can vary widely, causing problems in accu
until the patrons use the data and get results the
role should include evaluating the quality of

creasingly important function as the quantity continues to grow. Providing pointers to accurate sources of data on subject-specific Web pages helps to mitigate the problems found by simply surfing the Internet indiscriminately.

Finding foreign data is problematic at times. Data that are given to the public for free in the United States, such as data from geological surveys, must often be purchased from other countries. This is a source of frustration for the patrons, who then often come to the library to see if it is owned or if the library is willing to purchase these data. If other patrons will likely use the data and its purchase is consistent with the library's collection development policy, the library is often able to assist with these requests.

Very specific field data often are provided by colleagues rather than through library channels. It is not unusual for a patron to supplement these data with base map information from the library. For example, a patron doing fieldwork in Vietnam used a Global Positioning System (GPS) unit to locate her study sites, and then drew polygons representing land use categories upon satellite images in a GIS program. This type of gray data is often undocumented and not captured by the library's collection strategies. It therefore stays completely independent of the library and is accessible only to members of the specific project.

Hardware and Software

Libraries offering a full range of GIS services should provide their patrons with high-end computers, large screen monitors, and a color printer. Copious amounts of server space are needed to store large data sets and the necessary GIS programs. Branner Library houses four networked, 1.5 GHz Pentium PCs running Windows 2000 each with 40 GB of hard drive space. These computers are accompanied by 21-inch monitors, an HP color printer, a Contex 40-inch feed-through scanner, and a large-format HP plotter. The size of the computer lab will remain stable in the foreseeable future. GIS users at Stanford often prefer to work in their own offices, laboratories, or computer clusters. The library encourages patrons to use their own machines by providing easy access to online software and data, mitigating the need for the library to acquire additional computer resources.

Branner's Web site allows Stanford affiliates to download unlimited copies of GIS software and data to Stanford-owned machines. This free access is subsidized by the libraries' purchase of an annual site license with ESRI, currently the market leader for GIS software. In selecting GIS software for the library, Stanford University Libraries determined which software patrons most commonly used. ESRI makes popular GIS software such as ArcGIS and ArcView. Software has been downloaded from the site over 500 times to patrons in more than 30 different departments since the service was first offered.

Outreach

Campus outreach is an integral part of the lib der to increase awareness of the facilities and da tive outreach to different departments and grou regular basis through a series of workshops. The and a half and typically have three components: including a demonstration of the current ESRI data available through the library and on the Int discusses and demonstrates a relevant project. through departmental listservs, posters, and by v worked successfully for workshops targeting bi tions, the social sciences, and the humanities. U light the use of Census 2000 data in GIS and These workshops have increased the GIS user ba lated demand for library services.

GIS outreach is also done through in-class d strations are tailored specifically to the needs of sultation with the professor. In an Electrical E the case study below) students are given a long software to make them literate in its use, so the tool for subject-related analysis.

To introduce students from all disciplines to course is held yearly in the Department of Geolo ences. Branner's GIS staff leads one session. Th introduction to the software and shows the stud port USGS Digital Elevation Model (DEM) da focuses specifically on data, including how to fi tips for effective Web searching for data sets, an the quality of the data one finds.

CASE STUDIES

Electrical Engineering

One of the first groups on campus to express i ment of Electrical Engineering (EE). A professo the libraries for help using GIS software to create The objective was for students to create a compre lular telecommunication towers across a developi Knowing that GIS had the mapping and spatial

plish this goal, the library GIS staff worked with the professor to design a custom project for his students. (See Figure 1.) The teaching assistant (TA) for this course customized the software, ArcView, by adding tools that allowed one to place symbols representing radio towers on a country or province map. Designing this program with GIS software allowed the TA to include accurate spatial data representing man-made and natural features specific to each region. The customized project has been used for an assignment in the EE professor's course on rural telecommunication every fall quarter for the past five years. Each year, small adjustments are made to improve the application.

During one session each year, a GIS staff person from Branner Library gives a guest lecture introducing the fundamentals of GIS and ArcView software. This person then assists the TA with collecting and manipulating GIS data specific to each selected country or region into a project file. Most of these instructions are documented in a "TA Manual" that is used and updated by the TAs from year to year. Data on administrative districts, rivers, roads, populated areas, and digital elevation models are collected for the projects. These data are from ESRI Data

FIGURE 1. Map of a Cellular Telecommunications Network for Kyrgyzstan (Completed by a Student in Stanford's Electrical Engineering Department)

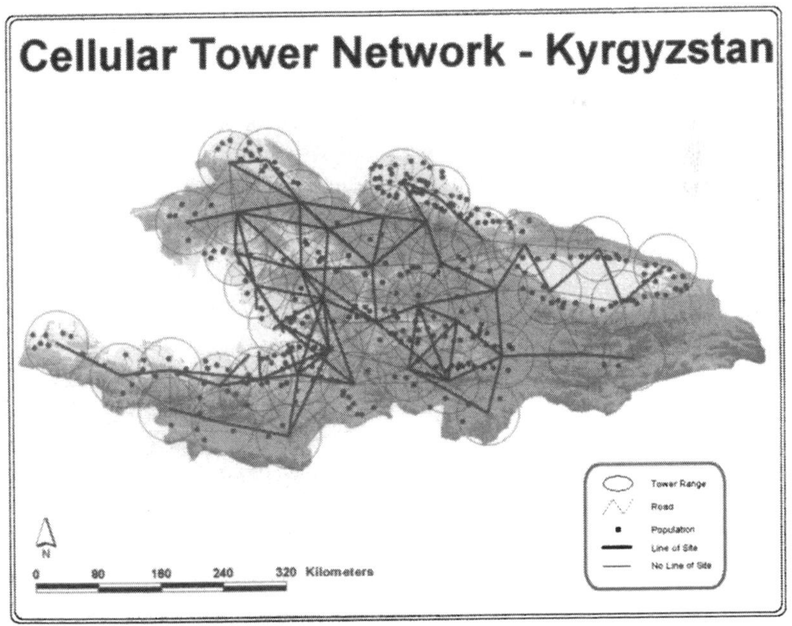

Reprinted with permission.

CD-ROMs in the library, and from the Internet, t
and GTOPO30[2] raster elevation data. If a project
than a whole country, the TA asks the staff for a
population data. The TA seeks data and technica
on behalf of his or her students, and this protoco
duration of the class's project. That is to say, the T
from the students and only consults GIS staff if
lem. Sometimes students will want to include add
ects, and the library staff will be consulted directl
search. More often the students accept the default
in completing the assignment, not developing a d
for an individual's research.

Students work together in small groups on the l
to complete their projects. Together they build a
by placing towers in locations where each tower n
neighbors. Tower locations are selected by consi
features such as rivers, roads, topography and to
consists of several maps as well as figures that co
sessing the cost of developing such a network. Stu
on these computers for the duration of the fall qu
of GIS gives the students an introduction to GIS
ation of a rural cellular telecommunication netwo
experience in a specific region, whereas before th
scription of the process of selecting tower locatio
rather sophisticated model without programming
that some professionals spend modeling the same

Epidemiology Research Group

In the past few years, researchers from the Ep
partment of Medicine at Stanford have begun u
group from the Stanford Center for Research in D
Marilyn A. Winkleby, Ph.D., was awarded a 5-ye
stitute of Environmental Health Sciences of the N
Their study was designed to assess the impact
physical environments on mortality, particularly
They began with the premise that neighborhood
and beyond their own individual characteristics
patterns and influencing behaviors partly through
services.

From the start, the study was designed to include the use of GIS in order to extend their empirical analysis, to study spatial patterns that would allow for the generation of new hypotheses, and allow for more meaningful communication of findings to the four study communities. According to Dr. Winkleby, GIS would enhance the way they interpreted the data. Spatial relationships would be much clearer when seen visually. GIS would allow them to ask different questions and to perform spatial analysis, which could augment more traditional statistical analyses.

In April 2001 the GIS & Map Librarian and the GIS Manager held a GIS workshop focusing on health research applications. Participation by researchers from the Department of Medicine and students in Epidemiology was high. After attending the workshop, Dr. Winkleby's group set up an individual meeting to find out more about the resources in the library and to ask for help in defining the parameters of the GIS project. They had scientists on the project with strong statistical analysis and database creation skills but only introductory GIS skills. It became clear that because of the specificity of their information needs, most data would not be readily available to them through the library. They would have to create it themselves. They needed exact addresses for the goods and services in the neighborhoods that might influence health. They conceptualized these as either assets (e.g., churches, grocery stores, parks) or barriers (e.g., alcohol distributors, fast food chains, crime).

The research group started with a database from a large, completed study conducted from 1979-1990, by the Stanford Heart Disease Prevention Program. This study focused on four Northern California cities: Modesto, Monterey, Salinas, and San Luis Obispo. Over 8,500 randomly selected male and female participants aged 12-74 were surveyed and assessed for cardiovascular disease risk factors from 1979-1990. The survey included information on socioeconomic status (education, income, occupation), cardiovascular disease risk factors (smoking, physical inactivity, high blood pressure, high cholesterol, obesity, and high dietary fat consumption), and other psychosocial and health-related risk factors. The current research project is using data for participants aged 25-74 from this original study. The survey and risk factor data were matched to death records through the end of 2000 to assess mortality outcomes. The addresses of the participants were geocoded and linked to census, archival, and other data in order to assess the neighborhoods' social and physical environments. (See Figure 2.) Based upon a recent study in the *American Journal of Public Health*, an excellent geocoding firm was hired to handle the geocoding of the data (Krieger et al., 2001). The resulting map layers were verified to test their spatial accuracy.

Recently completed, the first year of the study was heavily devoted to gathering data and creating the GIS using ArcView 3.2. The new neighborhood level

FIGURE 2. Distribution of Male and Female St
borhoods in Monterey, One of the Four Cities S
for Research in Disease Prevention

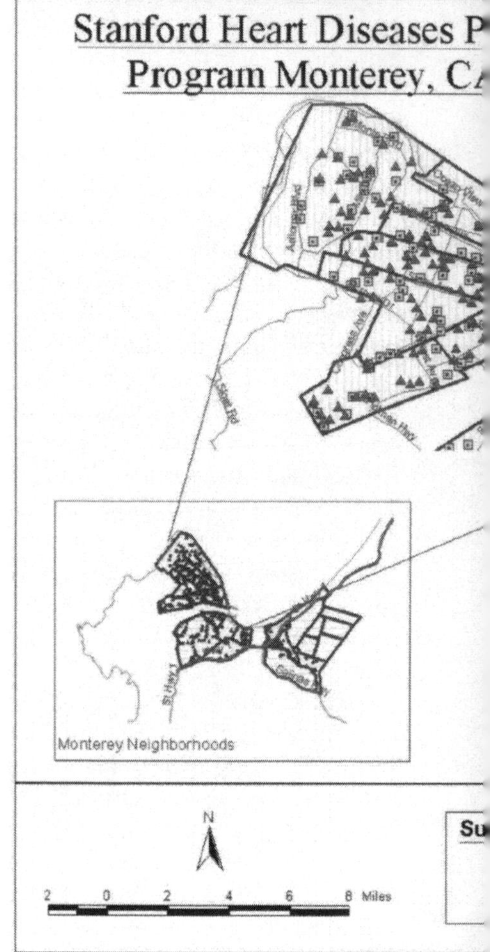

Reprinted with permission.

information needed was so complex and detaile
avenues to obtain it. The first task was to defin
from the study period of 1979-1990. Although t
Census boundaries, they also wanted to approxi
real neighborhoods as experienced by the partici

met with the researchers to help define neighborhoods that were meaningful to community residents. Then, GIS Census boundary files were borrowed from Stanford's Green and Branner libraries' data collections as well as downloaded from the Census Web site. Eighty-three neighborhoods were built by aggregating Census Block Groups. The geocoded addresses of the study participants were then associated with a neighborhood and linked to the survey data.

U.S. Census statistical data for 1980 (from the Census Bureau) and 1990 (from Wessex in SAS[4] format) were retrieved. Census data were used to calculate variables measuring neighborhood-level socioeconomic status, residential segregation, residential stability, family and age structure, and urban/rural status. These data were then linked to the neighborhood polygons in the GIS. Census population centroids from the 1990s, downloaded from the Internet, were employed to easily represent the neighborhood population in further analyses. For example, the centroids were used to calculate an average distance that the neighborhood's population must travel to certain goods or services, such as a park, pizza parlor, or primary care physician. Joining the socioeconomic Census information to the geographic areas built the social environment for each neighborhood.

After creating the neighborhoods, addresses for the neighborhood assets and barriers needed to be obtained. These data were wide-ranging and often required persistence and creativity to locate. Assets included parks, educational institutions and libraries, open spaces, banks and credit unions, houses of worship, youth organizations, day care centers, grocers and fresh fruit vendors, and primary care medical facilities. Barriers included liquor stores, pawnbrokers, tobacco shops, fast food restaurants, bars, and gun shops. GIS specialists in the four cities helped the researchers find relevant local data, and most offered use of the data without charge for the study. For example, parks were delineated with the help of GIS files, maps, and historical information provided by the city professionals. The majority of these sites were referenced by street address, and therefore required geocoding by the contracted firm for inclusion in the GIS. Once completed, the researchers drove by random sites to confirm the geocoded locations for quality control.

Telephone directories (white and yellow pages) often provided the best and most complete data for both assets and barriers, such as specific street addresses for stores, houses of worship, banks, and restaurants. These telephone directories were purchased from members of the community who came forward after ads were placed in local newspapers. Public libraries in Monterey and Salinas allowed the researchers to copy or scan addresses from telephone books at the library, but would not let the phone books circulate. Modesto telephone directories were purchased from Bell and Howell on microfiche. This part of the information-seeking process was the most laborious and time consuming. Data had to be manually entered into n then geocoded in the GIS. The California Scho Education Library at Stanford provided the mc and so were used in place of the phone books.

Alcohol licensing and accusation data (cita tions such as sales of alcohol to minors) were a California Department of Alcohol Beverage Co dress of each establishment selling alcohol, typ of the license. Through the use of a "key," accu tions, were accurately matched to a specific es detail was not attainable for tobacco sales. Neit hol, Tobacco and Firearms nor the state agencie of tobacco at a neighborhood level, nor is a lic

Crime data proved to be very problematic 1979-1990. "Uniform Crime Reports" during t addresses. On average, cities keep accurate crin of crimes for only 5 to 10 years. Police chiefs i and only one of the four, Salinas, had data that borhood level that fell within the timeline of th

Automobile collision statistics resulting in inj the National Highway Traffic Safety Administr (The California Highway Patrol had this informa data proved to be of limited value. In the study ci in any given year. In addition, street names wh corded, but not the cross streets or addresses, ma

Voting registration and participation record with the help of local County Clerk offices. The ically for the rosters for each presidential electic affiliation, address, and signature of voting men San Luis Obispo County and Monterey Count Monterey and Salinas) have agreed to release t

GIS zoning map layers for industrial section planning departments in each study area. It appe will be useable because zoning areas have remair

The process of collecting this information h ous scientists over the past year. It has called fc and university libraries; GIS specialists; local, s telephone books. It is the hope of the study's cr of data will be of use to many other researche health. With the data collection phase ending, t to expand the use of GIS in the project and to as multi-variate spatial analysis. The specialist has already been in contact with the library for software, data, and technical support. This research group exemplifies the variety of GIS needs that can be generated by one scientific project.

CONCLUSION

Libraries can and often do play a strong and influential role in the use of GIS on university campuses. By being a central point of contact for the acquisition of software and data, the library stays in touch with current research and the needs of the user community it serves. Outreach is critical in order to alert potential new users to the services available to them. Centralized, managed GIS support can save individuals and research teams a significant amount of time that would be spent obtaining site licenses, purchasing, organizing and distributing high quality data, and solving technical problems. Providing comprehensive Internet-based resources allows users to work independently in space and time by making the software and data available when the patron needs it, rather than when the library doors are open. GIS is an exciting, growing field and one in which the library can play a vital role throughout the campus.

NOTES

1. Digital Chart of the World data available at http://www.maproom.psu.edu/dcw/, September 30, 2002.

2. GTOPO30 data available at http://edcdaac.usgs.gov/gtopo30/gtopo30.html, September 30, 2002.

3. This work was co-funded by the National Institute of Environmental Sciences and the National Heart, Lung, and Blood Institute: Grant 1 RO1 HL67731 to Dr. Winkleby.

4. SAS Institute Inc.®–A statistical analysis system.

REFERENCES

Adler P.S., Larsgaard M.L. "Applying GIS in Libraries." In *Geographic Information Systems*. 2d ed. Vols. 2, *Management Issues and Applications*. New York: John Wiley & Sons, 1999.

Burrough P.A. "GIS and Geostatistics: Essential Partners for Spatial Analysis." *Environmental and Ecological Statistics* 8 (2001): 361-377.

Derksen, Charlotte R.M.; Juliet K Sweetkind; Meredith J. Williams. "The Place of Geographic Information System Services in a Geoscience Information Center." *Electronic Information Summit: New Developments and Their Impacts* in *Geoscience Information Society Proceedings* 31 (2000): 29-47.

Freyberg, Professor David, conversation with instruc
Kreiger, Nancy et al. "On the Wrong Side of the Tr
Geocoding in Public Health Research." *America*
(July 2001): 1114-1116.
Pinelli, Thomas E. "The Information Seeking Habits
ence & Technology Libraries. The Haworth Pres
Young, J.F. and L.C. Harriott. "The Changing Techn
cal Engineering 101:1 (January 1979): 20-24.

Interdisciplinary Research: A Literature-Based Examination of Disciplinary Intersections Using a Common Tool, Geographic Information System (GIS)

Robert S. Allen

SUMMARY. This study examines the interdisciplinary nature of articles that mention GIS or geographic information system in the title. Bibliographic information for 875 articles gathered from 6 years of an interdisciplinary database was used. Each article was coded based on a disciplinary classification scheme developed for the study. Articles were classified with either one or a number of disciplines assigned to each article. The data set was examined to determine the number of articles that belonged to each discipline. The interdisciplinary connectivity was examined by using single disciplines as focus disciplines, and measuring the percentage of overlap between the focus discipline and other disciplines. The disciplines that publish most in the area of GIS were determined. The interdisciplinary crossover between disciplines that publish in GIS was determined. *[Article copies available for a fee from The Haworth Document Delivery Service: 1-800-HAWORTH. E-mail address: <docdelivery@haworthpress.com> Website: <http://www.HaworthPress.com> © 2001 by The Haworth Press, Inc. All rights reserved.]*

Robert S. Allen, BS (Geology), MS (Geology), MS (Library & Information Science), is Head Librarian and Associate Professor of Library Administration, ACES (Agricultural Consumer and Environmental Sciences) Library, 1101 South Goodwin-MC 633, Urbana, IL 61801.

[Haworth co-indexing entry note]: "Interdisciplinary Research: A Literature-Based Examination of Disciplinary Intersections Using a Common Tool, Geographic Information System (GIS)." Allen, Robert S. Co-published simultaneously in *Science & Technology Libraries* (The Haworth Information Press, an imprint of The Haworth Press, Inc.) Vol. 21, No. 3/4, 2001, pp. 191-209; and: *Information and the Professional Scientist and Engineer* (ed: Virginia Baldwin, and Julie Hallmark) The Haworth Information Press, an imprint of The Haworth Press, Inc., 2001, pp. 191-209. Single or multiple copies of this article are available for a fee from The Haworth Document Delivery Service [1-800-HAWORTH, 9:00 a.m. - 5:00 p.m. (EST). E-mail address: docdelivery@haworthpress.com].

http://www.haworthpress.com/store/product.asp?sku=J122

10.1300/J122v21n03_12

KEYWORDS. GIS, geographic informa research, spatial data, bibliographic analys ture, engineering, hydrology, environmen

INTRODUCTION

Interdisciplinary research has been a growin for many years. There are a number of rapidly g many of the newest and most dynamic fields are i often mentioned example of one growing interdis tal science. This research field has shown its twentieth century. The subject developed by a from many existing disciplines to a broad theme broad theme increased, the new discipline emerg researchers interested in the broad theme applied ular disciplines, the nature of the interdisciplinar those new disciplines. There are many similar int developed along the same lines as the environm

The need to recognize developing interdisciplin ies when collections and services are being exami locked into collection and service patterns that are cipline parameters. It is necessary to understand th terdisciplinary areas, as these often are where the n place and where the new information sources are c cult for librarians to identify where the discipline new interdisciplinary fields. A better understandi search develops and how it can be defined is critical As the needs of various disciplines compete with th gets, the ability to asses the vitality and interdiscipl ing research fields is especially important.[1]

There are a number of recent articles in the lib light on interdisciplinary research. Many of thes techniques based on citation analysis to determine of scholarly writings. Citation analysis techniqu the interdisciplinary nature of research in physi forestry,[5] political science,[6] environmental geolog tronics,[9] marine sciences,[10] and the sciences in research has also been examined by survey techn tered to researchers.[13,14,15]

PURPOSE OF THE STUDY

The author of this article has long been interested in the implementation of GIS in both a university environment and in the working world outside the university. Implementing GIS services in libraries has been a popular undertaking during the past decade, with its most fruitful attempts being seen in the mid to late 1990s. The main impetus for this implementation interest was a project called the Association of Research Libraries GIS Literacy Project. This project served as the introduction of GIS technology and data to many librarians.[16]

There have been many assumptions made over the years about which disciplines actually employ GIS technology in their research efforts. Little concrete evidence has been provided that proves which disciplines are most likely to use this technology. Typically, if a library should choose to implement a GIS related service, then this service is often located within the map collection of said library. Map collections are often located in a departmental collection serving primarily geography, while some are located more within the confines of government documents. Sometimes map collections are located in other library locations, primarily based on their unique physical storage requirements. GIS services have also been implemented as a specialized reference service for university libraries or as part of a public library's service to the community.

The librarians most often involved with implementing GIS services in libraries, map librarians or GIS specialists, were often used to dealing with a limited group of library users. Greater knowledge of which disciplines within a research environment are most likely to use GIS technology will make it possible to reach out to those users not typically served by a library map collection. This study was initially intended to shed light on those disciplines that might not be expected to use map collections.

While carrying out the study, it became apparent that some important observations could be made about the over-lapping, or interdisciplinary, nature of the disciplines that write about GIS in their respective literatures. These observations begin to reveal some important considerations about different disciplines that employ a common tool to carry out their research. Though the common tool here is GIS, similar observations could be made in other interdisciplinary areas that use common tools. It would be similarly interesting, though perhaps less manageable, to investigate the use of a tool like nuclear magnetic resonance imaging for various purposes among disciplines. By examining the inter-related nature of the different disciplines classified in this study, some important observations can be made that describe how these disciplines might interact with each other.

METHODOLOG

A search was made of the combined subject
Editions. This database was made available witl
search terms consisted of the phrase "geograph
acronym "GIS." The search was limited to artic
further limited by requiring that the terms alread
article. The database was searched for publicatio
1999, and 2000. This yielded a dataset covering
that appear in all disciplines covered by Current
amined and any articles that did not deal with g
were removed from the set.

The dataset was then examined to determi
were included. The entire dataset was then coded
plines, most appropriate for each article. An i
many disciplines assigned to it as seemed appropr
one or a number of assigned disciplines. The en
to assess the disciplines and how they relate to ea
disciplines was examined in detail to determin
disciplines interacted with that discipline in the l

The author has many years of experience i
communities in many areas of science and tech
graduate degree in geoscience. This experie
enough level of knowledge in the associated a
accurate classification of the disciplines includ
The following list shows the disciplines include
inition of each discipline as determined by the

- *Engineering*–used when there is an obvi
neering, with the typical occurrence bein
- *Agriculture*–used when the article discuss
managed large scale food and crop produ
often distinguished from natural resourc
management applied.
- *Geoscience*–used when the article discuss
geology. Also used when the article discus
nature.
- *Environmental Science*–used when an a
events that have an effect on the environ

- *Biology/Ecology*–used when a specific naturally occurring plant or animal is discussed, or when the ecosystem for these plants and animals are a focus.
- *Urban/Regional Planning*–used when the subject deals with planning efforts in city management, or in large scale land planning on a regional basis.
- *Technology*–used when the article does not mention applying a GIS technique to a subject, but instead focuses on the technology of GIS without regards to a subject of application.
- *Natural Resources*–used when the article focuses on management of resources occurring in a natural environment. This is distinguished from ecology by the application of management techniques, with a product of potential interest to agriculture often being a focus.
- *Hydrology*–used when articles discuss hydrological techniques applied to movement of water, either in surface or sub-surface situations.
- *Geography*–typically used for articles that discuss the applications of GIS to topics in geography.
- *Library Science*–used for articles that deal with GIS services and data within a library setting.
- *Atmospheric Science*–used for articles that deal with study of the atmosphere and atmospheric conditions.
- *Education*–used for articles that pertain to educational uses of GIS, or in the education of GIS users.
- *Transportation*–used for articles that incorporate GIS into transportation planning activities.
- *Archeology*–used for articles that discuss the use of GIS in archeological research.
- *Criminology*–used for articles that incorporate GIS into the analysis of crime statistics.
- *Medicine*–used for articles that use GIS in planning medical services and tracking diseases.

RESULTS

There were a total of 875 GIS related articles in the six years of the database examined. The number of articles for each discipline, and the percent of the total for each discipline, is given in Table 1. The disciplines are ranked in the table to show in descending order the magnitude of publishing in GIS for each discipline.

TABLE 1. Number of Articles Coded for Each
for Each Discipline

Discipline	*Art*
Natural Resources	
Technology of GIS	
Urban Planning	
Environmental Science	
Biology/Ecology	
Hydrology	
Agriculture	
Engineering	
Geoscience	
Medical	
Transportation	
Geography	
Atmospheric Science	
Education	
Library Science	
Archaeology	
Landscape Architecture	
Criminology	
Total	

Table 2 shows the extent of interdisciplinary
These figures were based on the number of disc
There were 478 articles with only one subject disc
articles with two disciplines assigned. There were
assigned, and 9 articles with four disciplines assi
plines assigned for 397 articles, which made up

Table 3 provides a ranked list of the disciplin
disciplinary research found in each discipline.
the notion that articles having only one discipli
plinary than articles having multiple discipli
given for each discipline is based on the numb
coded within a single discipline. Hence, a disci
being single discipline would have 70% of its ar

Tables 4 through 14 provide the number of a
sect with a given focus discipline. Disciplines t

TABLE 2. Number of Disciplines Coded per Article

Disciplines per Article	*Articles*	*% of Total*
One Discipline	478	55%
Two Disciplines	309	35%
Three Disciplines	79	9%
Four Disciplines	9	1%
Multiple Disciplines	397	45%
Total Articles	875	100%

TABLE 3. Disciplines Ranked by Percent of Articles Within Discipline Coded as Single Discipline

Discipline	*Articles*	*% of Discipline*
Transportation	3	7%
Engineering	16	16%
Environmental Science	24	18%
Hydrology	24	21%
Agriculture	29	27%
Natural Resources	59	28%
Urban Planning	45	30%
Biology/Ecology	38	31%
Geoscience	31	34%
Medical	28	62%
Technology	110	67%

of articles were not included as tables for focus disciplines. The percentages given here are calculated using the total number of articles within a focus discipline and the number of articles that intersect that discipline. A percentage is also provided for the number of articles within a focus discipline that have no intersections, and these are termed as "single" in the tables.

DISCUSSION

The study results reveal that certain disciplines clearly have greater affinity for other particular disciplines. The data in Tables 4 through 14 indicate which disciplines intersect more often in the literature under examination. By choos-

TABLE 4. Number of Articles per Discipline that
as a Discipline

Discipline	Arti
Biology/Ecology	
Agriculture	
Hydrology	
Environmental Science	
Urban Planning	
Geoscience	
Technology of GIS	
Engineering	
Atmospheric Science	
Medical	
Transportation	
Education	
Library Science	
Landscape Architecture	
Criminology	
Archaeology	
Geography	
Natural Resources Single	
Natural Resources Total	

ing a focus discipline and looking at which disci
pline more often, an estimate of the affinity fc
with another discipline or disciplines can be m

The degree of interdisciplinary intersection
mated by looking at the data in Table 3. This tab
cles that do not have any intersecting discipl
amount of interdisciplinary research for a field.
rence of articles that were coded with only one d
cipline for an article would tend to yield an inte
less interdisciplinary than articles that are coded
This table would indicate that transportation is t
pline in the study, with only 7 percent of the arti
a single discipline. Engineering, environment
also strongly indicative of a high interdiscipli
nology are indicative of low interdisciplinary r
high percentage of single discipline coding.

TABLE 5. Number of Articles per Discipline that Intersect "Technology" as Focus Discipline

Discipline	***Articles***	**%**
Urban Planning	17	10%
Natural Resources	11	7%
Geoscience	7	4%
Biology/Ecology	6	4%
Agriculture	4	2%
Engineering	4	2%
Environmental Science	4	2%
Transportation	4	2%
Hydrology	3	2%
Geography	2	1%
Atmospheric Science	1	1%
Medical	1	1%
Library Science	1	1%
Education	0	0
Criminology	0	0
Archaeology	0	0
Landscape Architecture	0	0
Technology Single	110	67%
Technology Total	163	100%

A discussion of each separate discipline will examine how this data defines the interdisciplinary nature of the most commonly occurring disciplines in this study. The discussion will provide evidence of strong interdisciplinary linkages between disciplines. Likely reasons for why strong intersections occur are presented, and it is intended that such a discussion will provide greater knowledge for information professionals struggling to understand interdisciplinary research areas.

Geoscience: Geoscience articles in the study were most closely allied to articles coded as hydrology. The author expected this outcome, as hydrologists often study the movement of water from the surface into the subsurface. Water is also one of the strongest forces in shaping geological features. Civil engineering also intersects with geoscience a great deal, especially in the areas of rock and soil mechanics. The strong affinity for natural resources is based on articles that include a geological treatment of soil or water as resources.

TABLE 6. Number of Articles per Discipline that
Focus Discipline

Discipline	*Ar*
Transportation	
Engineering	
Environmental Science	
Technology of GIS	
Natural Resources	
Hydrology	
Biology/Ecology	
Agriculture	
Geoscience	
Medicine	
Education	
Atmospheric Science	
Library Science	
Landscape Architecture	
Criminology	
Archaeology	
Geography	
Urban Planning Single	
Urban Planning Total	

Hydrology: This discipline was most often as
natural resources. It was a bit surprising to see
paired with hydrology so often, but this is most lik
natural resource as well as being the basis for hy
rates complicated mathematics and modeling in it
to engineering was also expected. The importanc
hydrology is largely due to water being the me
many situations. Hydrology had a fairly high lev
tivity, probably due to its heavy reliance on theo
are applied in the disciplines mentioned as comm

Engineering: The articles coded for engineerin
disciplinary connectivity. This is most likely due t
gineers being applied to a variety of projects i
planning is the discipline most often connected to
ban planning is not usually considered an engine

TABLE 7. Number of Articles per Discipline that Intersect "Environmental Sciences" as a Discipline

Discipline	*Articles*	%
Agriculture	26	20%
Natural Resources	22	17%
Hydrology	20	15%
Engineering	18	14%
Urban Planning	15	14%
Biology/Ecology	13	10%
Medical	10	8%
Geoscience	8	6%
Transportation	7	5%
Atmospheric Science	7	5%
Technology of GIS	4	3%
Education	1	1%
Geography	0	0
Library Science	0	0
Landscape Architecture	0	0
Criminology	0	0
Archaeology	0	0
Environmental Science Single	24	18%
Environmental Science Total	131	100%

world applications, the engineering department of a city government often carries out much of the planning that goes on. Transportation research is also heavily connected to engineering. Hydrology also intersects often with engineering, especially in areas that attempt to model how water can be manipulated to bring about engineered outcomes. Environmental science was often connected to engineering, though it was expected that this figure might be higher prior to conducting the study due to the strong forces for environmental engineering in academia and private industry. Geoscience also showed a strong connection to engineering. Each of the disciplines mentioned as strongly connected to engineering have specialties of engineering professions associated with them. Examples are city engineers, hydrological engineers, transportation engineers, environmental engineers, or engineering geologists. It was expected that a higher connection to agriculture would be measured, due to the existence of agricultural engineering. This connection was not strongly evident in the study.

TABLE 8. Number of Articles per Discipline t
Focus Discipline

Discipline	
Natural Resources	
Environmental Science	
Urban Planning	
Technology of GIS	
Hydrology	
Atmospheric Science	
Agriculture	
Medical	
Transportation	
Geoscience	
Engineering	
Archaeology	
Library Science	
Education	
Landscape Architecture	
Criminology	
Geography	
Biology/Ecology Single	
Biology/Ecology Total	

Environmental Science: This is an example c
grown into a discipline in its own right. It was a
showed more interconnectivity than environm
pline did show a high degree of interdisciplina
connectivity with agriculture was also a bit of a
culture would cross over with environmental
was not expected at the level present in the lite
large amount of potential pollutants released in
izer, pesticide and agricultural waste is a major
disciplines. There was also a great deal of cros
drology and engineering. These connections are
eling through water. It was surprising that the
appear connected to environmental sciences m

Agriculture: Natural resources and enviror
two disciplines with the most cross over to a

TABLE 9. Number of Articles per Discipline that Intersect "Hydrology" as a Discipline

Discipline	*Number*	*%*
Geoscience	27	24%
Natural Resources	26	23%
Engineering	20	18%
Environmental Science	20	18%
Urban Planning	13	11%
Agriculture	9	8%
Atmospheric Science	4	4%
Biology/Ecology	4	4%
Technology of GIS	3	3%
Transportation	0	0
Library Science	0	0
Geography	0	0
Education	0	0
Medical	0	0
Landscape Architecture	0	0
Criminology	0	0
Archaeology	0	0
Hydrology Single	24	21%
Hydrology Total	114	100%

natural resources is based on the management of natural resources to create products of agricultural interest. These products are typically based in forests or fisheries. Soils also figure heavily into the cross over, as they are important to agriculture and are indeed an important natural resource. The connection with environmental science is often due to the introduction of pollutants into the environment through agricultural activities. It was expected that there would be a great deal of cross over with biology/ecology and engineering, but this was not evidenced to a large extent in the literature under study.

Natural Resources: This discipline showed the most connectedness to biology/ecology and agriculture. The cross over with biology/ecology is largely due to living natural resources, either plant or animal, being included in the discipline. If a considerable amount of management activity is applied to that resource to obtain a maximized product, then it becomes a part of agriculture. This particular three-way association forms a nice progression. The progression begins with the basic biological unit, and then continues on to the knowl-

TABLE 10. Number of Articles per Disciplin
Discipline

Discipline	A
Natural Resources	
Environmental Science	
Hydrology	
Atmospheric Science	
Urban Planning	
Technology of GIS	
Engineering	
Biology/Ecology	
Geoscience	
Medical	
Library Science	
Transportation	
Education	
Geography	
Landscape Architecture	
Criminology	
Archaeology	
Agriculture Single	
Agriculture Total	

edge that the biological unit is actually a natur and finally to extensive management and exp produce an agricultural commodity. Hydrol also show considerable connectivity to natur nection is primarily through the existence of the environmental science connection is thro ment factors that effect the environment.

Biology/Ecology: The only strong connecti resources, with nearly half of biology/ecolog plinary intersection. This is understandable, applying such a management tool to a biologi begin to indicate its importance as a natural re ology to natural resource to agriculture mentic when examining biology/ecology. There is l

TABLE 11. Number of Articles per Discipline that Intersect "Engineering" as a Discipline

Discipline	*Articles*	%
Urban Planning	30	31%
Transportation	20	21%
Hydrology	20	21%
Environmental Science	17	18%
Geoscience	16	16%
Natural Resources	9	9%
Technology of GIS	4	4%
Agriculture	4	4%
Atmospheric Science	2	2%
Biology/Ecology	1	1%
Library Science	0	0
Education	0	0
Medical	0	0
Landscape Architecture	0	0
Criminology	0	0
Archaeology	0	0
Geography	0	0
Engineering Single	16	16%
Engineering Total	97	100%

tween biology/ecology and agriculture. If the progression holds true, natural resources would be in between the two extremes.

Technology: The technology of GIS was the least interconnected discipline examined. In fact, there was little connection to other disciplines at all. The highest connections were shown for urban planning and natural resources; otherwise the connections with other disciplines were shown very lightly across most disciplines examined.

Transportation: This discipline showed the highest degree of interconnectivity. The disciplines with the highest degree of cross over to transportation were urban planning and engineering. The urban planning connection is largely due to the use of GIS to plan out traffic patterns and mass transit within cities. Engineering, while also involved in traffic management, is also heavily involved in planning out potential building sites for highways and railways.

Medical: The medical uses of GIS also showed little cross over to the other disciplines, with the exception being environmental science. This is

TABLE 12. Number of Articles per Discipline Discipline

Discipline
Hydrology
Engineering
Natural Resources
Environmental Science
Technology of GIS
Agriculture
Geography
Urban Planning
Biology/Ecology
Atmospheric Science
Transportation
Library Science
Education
Medical
Landscape Architecture
Criminology
Archaeology
Geoscience Single
Geoscience Total

most likely due to little connection in gene land based disciplines. The highest level of i ronmental science, and this proved to be the medical articles. The common use of GIS for vectors or examining the geographic distri events.

Geography: It is worth noting that very f any connection at all to geography. This may i database used to compile the data set. It may geographers may not be writing that much al that geography, as a pure discipline, has be graphical techniques in other disciplines. Th growth of interdisciplinary fields that formed existing discipline to a new and more vigoro

TABLE 13. Number of Articles per Discipline that Intersect "Medical" as Focus Discipline

Discipline	*Articles*	%
Environmental Science	10	22%
Biology/Ecology	4	9%
Transportation	3	7%
Natural Resources	2	4%
Urban Planning	2	4%
Technology of GIS	1	2%
Agriculture	1	2%
Geoscience	0	0
Hydrology	0	0
Library Science	0	0
Atmospheric Science	0	0
Education	0	0
Landscape Architecture	0	0
Criminology	0	0
Archaeology	0	0
Geography	0	0
Engineering	0	0
Medical Single	28	62%
Medical Total	45	100%

CONCLUSIONS

The results of this study were interesting for a number of reasons. First, it sheds light on which academic disciplines are most likely to use GIS. Second, it provides a dataset that can be effectively used to examine the interdisciplinary nature of land-based disciplines. Third, it can be used to educate librarians and GIS practitioners on common yet distinctly separate groups of GIS users that are evidenced in GIS literature.

This study is flawed for broad application beyond academia, as it does not delve outside the literature that is covered within the limited database used for the sample. It is common knowledge that GIS is used for a number of purposes that did not surface in the literature set examined. It also does not effectively measure which disciplines use GIS more, only which disciplines write about GIS more.

The dataset examined is also limited to the literature included within the database used and to the size of the literatures of each respective discipline. For a

TABLE 14. Number of Articles per Discipline
Focus Discipline

Discipline	
Urban Planning	
Engineering	
Environmental Science	
Technology of GIS	
Medical	
Transportation Single	
Biology/Ecology	
Natural Resources	
Atmospheric Science	
Agriculture	
Library Science	
Hydrology	
Education	
Geoscience	
Landscape Architecture	
Criminology	
Archaeology	
Geography	
Transportation Single	
Transportation Total	

more complete study, it would be necessary t discipline-oriented databases to compile a c could then be weighted based on the number base and dataset for each discipline.

REFERENCES

1. Palmer, C.L., editor. 1998. *Navigating amc interdisciplinary inquiry*. Published as a special iss *Fall 1996*. Champaign: University of Illinois at U of Library and Information Science, 1996, 238 pag

2. Rinia, E.J. et al. 2002. "Impact measures of ics." *Scientometrics*. Vol. 53, No.2, p. 241-8.

3. Hrubel, J.V.M. and E.A. Goedeken. 2001. Citations Index to identify a community of interdis bibliometric study." *The Serials Librarian*. Vol. 41

4. Schwechheimer, H. and M. Winterhager. 2001. "Mapping interdisciplinary research fronts in neuroscience: a bibliometric view to retrograde amnesia." *Scientometrics*. Vol. 51, No. 1, p. 311-18.

5. Steele, T.W. and J.C. Stier. 2000. "The impact of interdisciplinary research in the environmental sciences: a forestry case study." *Journal of the American Society for Information Science*. Vol. 51, No. 5, p. 476-84.

6. Pierce, S.J. 1999. "Boundary crossing in research literatures as a means of interdisciplinary information transfer." *Journal of the American Society for Information Science*. Vol. 50, No. 3, p. 271-9.

7. Zipp, L.S. 1999. "Core serial titles in an interdisciplinary field: the case of environmental geology." *Library Resources and Technical Services*. Vol. 43, No. 1, p. 28-36.

8. McCain, K.W. 1998. "Neural networks research in context: a longitudinal journal cocitation analysis of an emerging interdisciplinary field." *Scientometrics*. Vol. 41, No. 3, p. 389-410.

9. Hinze, S. 1994. "Bibliographical cartography of an emerging interdisciplinary discipline: the case of bioelectronics." *Scientometrics*. Vol. 29, No. 3, p. 353-76.

10. McCain, K.W. 1992. "Core journal networks and cocitation maps in the marine sciences: tools for information management in interdisciplinary research." *Proceedings of the 55th Annual Meeting of the American Society for Information Science, Pittsburgh, Oct. 26-29, 1992*. p. 3-7.

11. Qin, J., Lancaster, F.W., and B. Allen. 1997. "Types and level of collaboration in interdisciplinary research in the sciences." *Journal of the American Society for Information Science*. Vol. 48, No. 10, p. 893-916.

12. Hurd, J.M. 1992. "Interdisciplinary research in the sciences: implications for library organization." *College and Research Libraries*. Vol. 53, No. 4, p. 283-97.

13. Spanner, D. 2001. "Border crossings: understanding the cultural and informational dilemmas of interdisciplinary scholars." *Journal of Academic Librarianship*. Vol. 27, No. 5, p. 352-60.

14. Palmer, C.L. 1999. "Structures and strategies of interdisciplinary science." *Journal of the American Society for Information Science*. Vol. 50, No. 3, p. 242-53.

15. Qin, J., Lancaster, F.W. and B. Allen. 1997. "Types and levels of collaboration in interdisciplinary research in the sciences." *Journal of the American Society for Information Science*. Vol. 48, No. 10, p. 893-916.

16. Adler, P.S. 1995. "Special issue of geographic information systems (GIS) and libraries: an introduction." *Journal of Academic Librarianship*. Vol. 21, No. 4, p. 231-32.

Index

Numbers followed by *fig*. indicate figures; those followed by *n* or *nn* indicate note(s).

About, Web directory guide, 28
Acoustic Material Property Tables, 40, 41*table*
A2D, 53
Adams, Mignon, 2,87
Advanced Thermal Analysis Laboratory, 40
African Mammals Databank (Institute of Applied Ecology), 108
AGI (American Geologic Institute), 56
AGRICOLA database, 106,124
Agriculture academic discipline. *See* Geographic Information System (GIS), interdisciplinary research using
AHFS Drug Information, 93,98
Albany International Research Co. *See* Current awareness (CA) reports at Albany International Research Co.
Allen, Robert, 3,191
American Geologic Institute (AGI), 56
American Zoo and Aquarium Association (AZA), 103,111
Archaeology academic discipline. *See* Geographic Information System (GIS), interdisciplinary research using
Arizona State University Index to Physical, Chemical, and Other Property Data, 42,42*table*
ARS Pesticide Properties Database, 38,38*table*
Association of Research Libraries GIS Literacy Project, 193
ATHAS Data Bank of Thermal Properties, 40,41*table*
Atmospheric science academic discipline. *See* Geographic Information System (GIS), interdisciplinary research using
AZA (American Zoo and Aquarium Association), 103,111

Beilstein Crossfire database, 9,11,13,14, 25*n. 10*,28,31,35-36,35*table*
Best Evidence medical article database, 93
Bibliographic sources. *See* Hydrology online bibliographic sources
Biblioline (Wildlife & Ecology Studies Worldwide), 105
Biochemists. *See* Chemistry faculty
Biological Abstracts, 124
Biological Resources Discipline (BRD), of USGS, 67
Biology academic discipline. *See* Geographic Information System (GIS), interdisciplinary research using
BioMed Central, 14
BioOne, 105,106
BIOSIS, 124

http://www.haworthpress.com/store/product.asp?sku=J122

Biosis Previews, 105,106
Botanical information, resources and user needs, 2
  collaboration, 123,126
  electronic resources, 123-125
    databases, 124
    e-journals, 123
    georeferenced maps, 123-124
    GIS, 123
    subscription-based online resources, 125
  information-seeking behavior, 122-123
  institutes and programs, 127-128
  interdisciplinary focus, 121
  printed resources, 125-126
  search tools, 122
  summary regarding, 121
  unpublished literature, 126
Botanico-Peridicum-Huntianum, 125
Branner Earth Sciences Library and Map Collections (Stanford University). *See* Geographic Information System (GIS), academic library user information needs

CA. *See* Current awareness (CA) reports at Albany International Research Co.
CAB abstracts, 124
CambridgeSoft, 32-33
Captive Breeding Specialist Group (CBSG), 103,112
Caracuzzo, Alex, 3,165
CATALPA, 124
CBHL (Council of Botanical and Horticultural Libraries), 126
CBSG (Captive Breeding Specialist Group), 103,112
Center for Information and Numerical Data Analysis and Synthesis, 29

study interview questions, 10-11
study interview results
access improvement, 5
convenience, 5
electronic journals, 19-24
faculty roles, 11-12,25*n. 9*
future directions, 21-22
keeping current, 15-19
self-reliance, 5
sophisticated tool selection, 5
summary regarding, 5,22-24, 24*n. 22*
time-savings, 5
tools of the trade, 12-14,15*table*, 25*nn. 10,11*
study limitations, 5
study methodology, 10-11
subfield differences, 23
survey methodology, 5-6
TOC scanning, 16
at University of Texas, Austin, 9-10
See also Chemical physical property information web sites
Chimpanzee Cultures Web site, 109
CiteSeer, 14
Coates, Linda, 2,101
Cochrane Database of Systematic Reviews medical database, 93
Cochrane Library Online, 98-99
Colby College, 37
Committee on the Preservation of Geoscience Data and Collections, 59
Company Name Finder DIALOG database, 170
Comparative Placentation Web site, 110
Consultant vet diagnostic Web site, 110
CORE e-journal project, Cornell University, 19
Council of Botanical and Horticultural Libraries (CBHL), 126
CRC Chemical Dictionaries on CD-ROM, 125
CRES (Center for Reproduction of Endangered Species), 102
Criminology. *See* Geographic Information System (GIS), interdisciplinary research using
Critical Properties of Gases, 38,38*table*
Cullman Program for Molecular Systematic Studies, 127-128
Current Awareness (CA) reports at Albany International Research Co.
challenges for
PMS database shortage, 168
textile information abundance, 168
content vendors, 170
filtering and formatting, 170-171
future of, 173
history of, 166-167
information user differences, 167
marketing of, 171-172
measuring success of, 172
R&D use of, 168-169
competitive intelligence, 169-170
DIALINDEX database, 170
DIALOG database, 170
new product ideas, 169
supporting projects, 169
summary regarding, 165
Current Contents TOC services, 16-17

Design Institute for Physical Property Data, 29
DIALINDEX database, 170
DIALOG database, 170
Dictionary of Organic Compounds, 29
Dielectric Constant Reference Guide, 40-41,41*table*
Drug Facts and Comparisons, 93,98
Drugdex, 93
Duke University Chemical & Physical Properties in the Library, 42,42*table*

E-journals. *See* Electronic journals
Ecology academic discipline. *See* Geographic Information System (GIS), interdisciplinary research using

Education academic discipline. *See* Geographic Information System (GIS), interdisciplinary research using
Electrical engineering GIS case study, 182-184,183*fig.*
Electronic journals, 19
- archival issues, 2
- of botanists, 122-123
- chemistry, 19-21
- cost issues, 22
- hard copy preferences, 19-20,24
- hypertextual crosslinking feature, 21-22
- of petroleum geologists, 50
- publisher ethics, 22
- time savings, 19-20
- TOC scanning, 16,17,92,98,113

Electronic newsletters, 98
Embase biomedical science database, 92,97
EMBL Reptile Database, 109
EndNote bibliographic software, 25-26*n. 11*
Engineering. *See* Engineering knowledge community; Geographic Information System (GIS), interdisciplinary research using
Engineering knowledge community, 3
- engineering knowledge, nature of, 145-147
  - anatomy of design knowledge, 146
  - cognitive activity types, 146
  - "demarcationist" concept, 146-147
  - descriptive knowledge, 145-146
  - explicit *vs.* implicit, 146
  - network actor model, 147
  - prescriptive knowledge, 145-146
  - tacit knowledge, 146
  - variation-selection model, 146
- engineers, aerospace community, 131,155-157
  - community norms, 157

engineers using interactive parameter variation, selective retention, 144-145
scientists generating new knowledge, 144
scientists using hypothesis testing, 144-145
engineers and scientists, similarities, 141
knowledge generation methods, 140
technological change methods, 141
science *vs.* technology, 131,133-137
actor and societal levels of, 134-135
as consumers of information, 136
continuous relationship of, 135
differences between, 135-136
engineers transforming information, 136-137
independent nature of, 136
inputs *vs.* outputs of, 136-137
rule of priority of science, 133-134
rule of secrecy of technology, 134
science, introverted activity, 133
science base of technology concept, 132
as social organizations, 133
technological advance concepts, 132
technology, extroverted activity, 133
technology, inputs *vs.* outputs of, 137
scientific and technological communities
community sociologies, 142-143
engineering "college" *vs.* scientific "knowledge community," 142
engineering diversity, 143
engineering information exchange, 143
engineering interpersonal communication, 143-144
"inside" *vs.* "outside" orientations, 142
structured social settings, 141-142
summary regarding, 131,157-159
Environmental science
GIS usage, 3
See also Geographic Information System (GIS), interdisciplinary research using
Environmental Systems Research Institute (ESRI) Data & Maps CD-ROM, 179
EOI Suite (Estimation Program Interface), 36
Epidemiology GIS case study, 184-189, 186*fig.*
Estimation Program Interface (EOI Suite), 36

Faculty. *See* Chemistry faculty
Faculty of 1000, 14
Flaxbart, David, 2,5
Flora Brasiliensis, 126
Flora of Guatemala, 126
Flora of Peru, 126
Fraser, Susan, 2,121
Fuel Property Database, 38-39,38*table*

Geographic Information System (GIS), academic library user information needs and, 3,176
access issues, 177
case studies
electrical engineering, 182-184, 183*fig.*
epidemiology, 184-189,186*fig.*
description of, 177-178
GIS support at Branner
data, 179-181
hardware and software, 181

search strategies, 180-181
tutorials, 178-179
workshops, demonstrations, 182
historic background regarding, 176-177
importance of, 176
spatial data focus, uses of, 176
summary regarding, 175
See also Geographic Information System (GIS), interdisciplinary research using
Geographic Information System (GIS), interdisciplinary research using, 1,3,176-177
Association of Research Libraries GIS Literacy Project, 193
botanists, 123
citation analysis, 192
interdisciplinary fields recognition, 192
petroleum industry, 48
study discussion, discipline intersections, 196*table,* 197-199,197*tables*
agriculture, 194,202-203,204*table*
archaeology, 195
atmospheric science, 195
biology, ecology, 195, 202*table,*204-205
criminology, 195
education, 195
engineering, 194,200-201,205*table*
environmental science, 192,194, 201*table,*202
geography, 195,206
geoscience, 194,199,206*table*
hydrology, 195,200,203*table*
library science, 195
medicine, 195,205-206,207*table*
natural resources, 195,198*table,* 203-204
technology, 195,199*table,*205
transportation, 195,205,208*table*
urban planning, 195,200*table*

Handbook of Environmental Data on Organic Chemicals, 29
Harrison's Principles of Internal Medicine, 94,99
Hart's E&P, 59
Havener, Michael, 2,63
Hazardous Substances Data Bank, 36-37
Heard's Zoological Restraint & Anesthesia Web site, 111
Herblit Database, 106
High-Temperature Superconductors (WebHTS), 39
HORT CD, 124
Hydrology academic discipline. *See* Geographic Information System (GIS), interdisciplinary research using
Hydrology online bibliographic sources
  collaboration, 66
  commercial bibliographic databases, 78,80-83
  GIS usage, 3
  government information sources, 2
  gray literature, 66-67
  ground-water hydrologists, defined, 64
  hydrology, defined, 64
  importance of, 66-67
  non-commercial bibliographic databases, 78,80-81,83-84
  project deadline issues, 66-67
  saltwater-intrusion assessment
    commercial bibliographic database, 81-83
    non-commercial bibliographic database, 83-84
    overview, 64-65,65*fig.*
    selected keywords and subject headings, 78-80*tables*
    state geological survey information sources, 75,76-77*tables*
  state and other federal sources
    academic and government library catalogs, 75-78,77*table*
    state geological survey information sources, 75,76-77*tables*
    terminology variations, 78-80*tables*
    water-resources agencies information sources, 75,76-77*tables*
    WRD state office bibliographic sources, 68-69*tables,*74
  summary regarding, 63,84-85
  surface-water hydrologists, defined, 64
  U.S. Geological Survey sources
    alternative sources from, 74-75
    Atlantic coastal zone web sites, 68-69*tables*
    availability and access issues, 68, 70-71,70*fig.*
    distribution structure of, 68,70-74
    national information sources, 71,72-73*tables,*73-74
    overview of, 67-68
  See also Geographic Information System (GIS), interdisciplinary research using

IBGE Cidades@, 124
IEB (Institute for Economic Botany), 127
IHS AccuMap, Ltd., 54
IHS Energy Group, 53,54,59
Index Herbariorum, 124
Index Londinensis (Royal Horticultural Society of London), 125-126
Index Nominum Genericorum, 124
The Index to American Botanical Literature, 124
Indiana University, 28
Indiana University CHEMINFO SIRCh Physical Properties, 42*table,*43
Institute for Economic Botany (IEB), 127

Institute of Systematic Botany (ISB), 127
Intermountain Flora, 126
International Chemical Safety Cards, 37
International Code of Botanical Nomenclature, 124
International Critical Tables, 28
International Pharmaceutical Abstracts, 97
International Plant Names Index (IPNI), 124
International Rhino Foundation Web site, 109
International Union for Conservation of Nature and Natural Resources, 111
International Working Group on Taxonomic Databases for the Plant Sciences (TDWG), 123
Internet resources. *See* Chemical physical property information web sites; Hydrology online bibliographic sources; Pharmacist information needs; San Diego Zoo Library
Internet Scout Project, University of Wisconsin, 28
Invisible Web, 28
IPNI (International Plant Names Index), 124
ISB (Institute of Systematic Botany), 127
ISIS pedigree and breeding records, 108
IUCN Cat Specialist Group, 109
IUCN Red List of Threatened Species, 112,124

Joseph, Laura, 2,47
Journal of Chemical and Engineering Data, 29
Journal of Physical and Chemical Reference Data, 29
JSTOR, 123

King, Donald, 5

La
Lit
Lo
Lu
Ma
Ma
Ma
Ma
Mc
ML
Me
Me
Me
Mic
MJ
MS
MS
Nat
Nat
Nat
Nat
Nat

National Research Council, 59
National Toxicology Program (NTP), 37
Natural resources academic discipline. *See* Geographic Information System (GIS), interdisciplinary research using
NCMS (National Center for Manufacturing Sciences), 40
NEC ResearchIndex, 14
New York State Department of Environmental Conservation, 54
NIH-NHGRI genome database, 14
NIMA (National Imagery and Mapping Agency), 180
NISC (National Information Services Corporation), 105
NIST Ceramics WebBook, 38*table,*39
NIST Chemistry WebBook, 32*table,* 33,35,35*table,*39
NIST (National Institute of Standards and Technology), 33
North Dakota Industrial Commission Oil and Gas Division, 54
NTP Chemical Health & Safety Data, 37
NTP (National Toxicology Program), 37

Oak Ridge National Laboratory, 40
OCLC (Online Computer Library Center), 122,124
Oil and Gas Journal, 59
Online bibliographic sources. *See* Hydrology online bibliographic sources
Online Computer Library Center (OCLC), 122,124
Open Directory Project, 28
Organic chemists. *See* Chemistry faculty
Organic Compounds Database, 37
The Origins of the Turbojet Revolution (Constant), 157

Pesticide Fact Sheets–New Active Ingredients, 41,41*table*
Petroleum Abstracts Bulletin, 59
Petroleum geologists, information-seeking, communication behavior of, 2
  digital environment conversion, 47, 48-51,60-61
    3-D immersive visualization, 48-49
    Geographic Information Systems tool, 48
  engineers, researchers *vs.,* 51
  information storage, 50
  information storage changes, 50
  information threats, 59-60
  information transfer changes, 2,49
  journal literature access, 50
  proprietary information, 2,49,53
  recent changes in, 48-51
  seismic data, 50-51
  summary regarding, 47-48,60-61
  task: creating prospects
    gray literature sources, 56,59
    information resources used, 52-55
    legal spacing requirements knowledge, 54-55
    literature searches, 55-56
    maps and cross section tools, 54, 56-57
    mud logs, 52-53,58-59
    process of, 55-58
    production information, 54,56
    sample logs, 52-53
    scout cards, 52,53-54
    well logs, 52
    wireline logs, 53
  task: industry activity, keeping current with, 59
  task: oil and gas fields development, 59
  task: well site work, 58-59
Pharmacist information needs, 2
  drug formulary development, 90,93
  educational needs, 88,94-95

- employment opportunities, 88
- of hospital and consultant pharmacists, 90-91
- information gathering methods
  - article review, 92-93
  - clinical studies, 91-92
  - drug monographs, 93
  - practice guidelines, 93
  - textbooks, 93-94
  - websites, 93-94
- librarian roles
  - database use, 95-96
  - medical literature access, 95-96
  - *Medline* search features, 95
  - search question formulation, 95
- of pharmaceutical industry pharmacists, 91
- resources
  - drug monographs, 98
  - indexes and abstracts, 97
  - newsletters and alerting services, 97-98
  - systematic reviews, meta-analyses, 98-99
  - websites, 96-97
- of retail, community pharmacists, 89-90
- summary regarding, 87,96

Phase Diagrams Web, 41,41*table*
Physical chemists. *See* Chemical physical property information web sites; Chemistry faculty
Physical Properties Database (PHYSPROP), 32*table*,34, 35,35*table*
Physical properties of chemicals. *See* Chemical physical property information web sites
Physician's Desk Reference, 93
PHYSPROP (Physical Properties Database), 32*table*,34,35*table*
PI/Dwights PLUS Energy News On Demand, 59
PI/Dwights PLUS Production Data, 54
Pinelli, Thomas, 3,131

IUCN, 111-112
Species Survival Commission (SSC), 111-112
geographic distribution of, 101, 104-105
information resources of, 2
interlibrary loan, 113-115
library *vs.* zoo collections, 103-104
summary regarding, 101-102
survey follow-up, 116-118
survey of scientific staff
results, 118-120
search tools, 105-106
veterinary medicine information, 109-110
Comparative Placentation, 110
Consultant diagnostic tool, 110
Heard's Zoological Restraint & Anesthesia, 111
McKenzie's Capture & Care Manual, 111
Merck Veterinary Manual, 110
university faculty collaboration, 102
WildPro information network, 110
virtual reference, 115-116
Web-based intranet portal of, 101-102,104-105
web resources of, 106-107
Zoological Society News, 112-113
Scholar database, 9
Science Citation Index, 13
Science Direct, 14
Scientists. *See* Engineering knowledge community
SciFinder database, 9,11,13,14,17,23, 24,25*n. 10*,25*n. 11*,28,170
SciFinder Scholar, 28
Solv-DB, 38*table,* 39-40
Specialty Information Associates, 41
Species Survival Commission (SSC), 103,111-112
SSC (Species Survival Commission), 103,111-112
Standard Reference Data Program Publications, 28
Stanford University Library System. *See* Geographic Information System (GIS), academic library user information needs
StatRef!, 99
STN® International Databases, 28,30,56
Structural Ceramics (WebSCD), 39
SUMSearch medical database, 93,99
Sweetkind-Singer, Julie, 3,175
Syracuse Research Corporation, 34
Syracuse Research Institute, 36

Taxonomic Literature 2 (Stafleu, Cowan), 125
TDWG (International Working Group on Taxonomic Databases for the Plant Sciences), 123
Technology academic discipline. *See* Geographic Information System (GIS), interdisciplinary research using
Tenopir, Carol, 5
Texas Railroad Commission, 55
Thermodynamics Research Center, 29
3-D immersive visualization, 48-49
Transportation academic discipline. *See* Geographic Information System (GIS), interdisciplinary research using
Trinity University, 28
TRIP (Turning Research into Practice) medical database, 93,99
TROPICOS, 124
Turning Research into Practice (TRIP) medical database, 93,99

Ultimate Ungulate Web site, 109
UnCover TOC services, 16-17
University at Buffalo, 28
University at Buffalo Materials Properties Locator Database, 42*table,*43

University of Akron, 36
University of Chicago, 28
University of Texas, Austin. *See* Chemistry faculty
University of Texas Thermodex, 42*table,*43
Urban planning academic discipline. *See* Geographic Information System (GIS), interdisciplinary research using
U.S. Geological Survey. *See* Hydrology online bibliographic sources

Vanderbilt University, 28
Vanderbilt University Finding Chemical & Physical Properties, 42*table,*43
Virtual reality visualization systems, 49

Wagner, A. Ben, 2,27
Walker's Mammals of the World, 105
Water Resources Discipline (WRD), of USGS, 68
Web of Science database, 13,14
Web resources. *See* Chemical physical property information web sites; Hydrology online bibliographic sources; Pharmacist information needs; San Diego Zoo Library

Printed and bound by CPI Group (UK) Ltd, Croydon, CR0 4YY
17/10/2024
01775687-0004